D0025199

Robotics and Automated Manufacturing

Richard C. Dorf

Reston Publishing Company, Inc.
A Prentice-Hall Company
Reston, Virginia

Library of Congress Cataloging in Publication Data
Dorf, Richard C.
  Robotics and automated manufacturing.
  Bibliography: p.
  Includes index.
  1. Robots, Industrial   I. Title.
TS191.8.D67   1983   629.8′92   82-23134
ISBN 0-8359-6686-0

© 1983 by Reston Publishing Company, Inc.
A Prentice-Hall Company
Reston, Virginia 22090

10   9   8   7   6   5   4   3   2

Printed in the United States of America

*To my wife, Joy,
who has provided
encouragement, understanding,
and helpful reflection on this subject.*

Contents

Figures

Tables

Preface

The continued prosperity of our nation depends upon the utilization of technological advances in the production of goods and services. The challenge of international competition and domestic industrial obsolescence requires that our genius for technological innovation become a central concept in this decade.

The use of robots and computer-aided design and manufacturing systems throughout the industries of the United States will be an increasing factor in the maintenance of our industrial leadership in the world. The purpose of this book is to consider the fundamental concepts and applications of robots and computer-aided manufacturing systems that may be effectively utilized in the nation's manufacturing plants and diverse work places. The development of robots and automated systems for mining, space, undersea, office, and manufacturing applications will be of central importance to the improvement of the nation's productivity.

This book is written for the engineer working in industry or government. It should be of value to engineers from many disciplines, since the field of robotics and automation is widely applicable and multidisciplinary. The book will also be useful as a textbook in university courses and for continuing education.

The book first considers the issues of productivity and automation. It then reviews the history, development and classification of robots, their mechanical and electrical components, and sensors and vision systems. Computers and artificial intelligence are discussed, and the wide range of applications of robots are presented. Economic, labor, and work place considerations are provided, and finally the future of robotics and automated manufacturing is considered.

This book incorporates the ideas and concepts of many engineers and scientists. I appreciate the assistance provided by them, as well as by the many firms active in the robotics industry.

Richard C. Dorf
University of California, Davis

Productivity and Industrial Competitiveness

The United States of America, the world's leading industrial giant for nearly a century, is now only one nation among many competing for the world's markets for manufactured goods. This worldwide challenge is compounded by a lag in the nation's productivity growth and increasing inflationary pressures. Flagging productivity has been diagnosed by many as the new American industrial disease, and its symptoms and causes have become an issue of grave concern to the nation's governmental and industrial leaders.

Productivity is defined as output per worker-hour, usually expressed in units manufactured per worker-hour invested. Using this measure of labor productivity, the rate of growth of production in the U.S. has slowed down significantly over the past decade, as shown in Table 1-1. The nation's average rate of productivity growth was 3.2% for 19 years following World War II, but dropped to an average of 1% for the period 1973–1978 and actually declined during the years 1978 through 1980. In addition, United States productivity has not kept pace with the manufacturing productivity growth of other industrial nations, as shown in Figure 1-1.

Prospects for a simple solution to the nation's flagging productivity, however, are not easily obtained. Those studying the problems see it as a complex phenomenon with multiple causes. During a period of inflation, the pressure for wage increases is strong and costs of production go up. Without a rise in productivity, businesses can meet higher costs only by cutting profits or raising prices. All of this exacerbates inflation and puts heavy pressure on the dollar.

TABLE 1-1.   U.S. Productivity, Output per Worker-Hour

Period	Average Rate of Growth
1947–1966	3.2%/year
1966–1973	2.1%/year
1973–1978	1.0%/year

Year	Growth for the Year
1978	−.2%
1979	−.7%
1980	−.3%
1981	+.9%

Note: This table records total productivity
growth for all sectors of the U.S. economy.

America's relatively high productivity rate in the years immediately following World War II owed much to the movement of agricultural workers to other sectors of the economy, especially the service industries. Agricultural workers are now such a small part of the work force that this trend can no longer be considered a significant source of gain in productivity. Also, the relatively high educational level of the nation's work force is no longer viewed as advantageous relative to other industrialized countries.

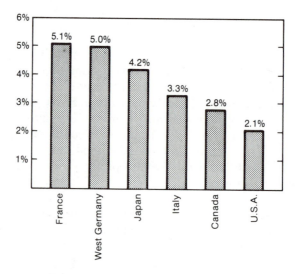

FIGURE 1-1.   Average manufacturing productivity growth 1973–79. The U.S. lags behind other industrialized countries. *Source: U.S. Department of Labor, Bureau of Labor Statistics.*

Some economists assign partial blame for the slump in productivity to a shift in the nation's work force to the service sector, which includes wholesale and retail trade, finance, insurance, real estate, the professions, business services, and general government (see Figure 1-2). The percentage of the labor force in this sector rose from about 50% to 60% between 1950 and 1960, and has now reached a level of 70% of the total employment of the nation. America is rapidly becoming a white-collar bureaucratic nation, partly because our educational system is supplying large numbers of new workers who are trained to work in the service economy.

Despite the increase of workers in the service sector, studies have shown that productivity in that sector has risen more slowly in most years than in the industrial sector, which includes manufacturing, mining, construction, communications, public utilities, and certain financial enterprises (Flint, 1981). Manufacturing employment has dropped to fewer than 14 million people, and has reduced nearly one million during the last 15 years; yet productivity in this sector increased about

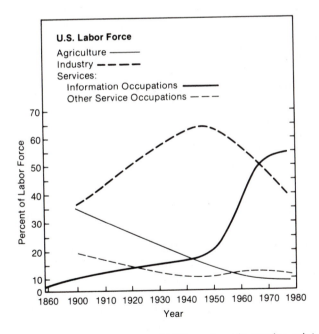

FIGURE 1-2.   The percentage of U.S. workers in service-related industries has been increasing steadily since 1960, with the most dramatic increase beginning in the 1950's. *Source: U.S. Congressional Committee on House Administration.*

2% during the period 1973 to 1978, or about twice the overall national average of 1% per year shown in Table 1-1. This is largely due to technological advances in automated manufacturing.

Technology and Productivity

The focus of this book is on the manufacturing sector, specifically the introduction of automation and other capital-intensive approaches to improving productivity. The Office of Technology Assessment of the Congress of the United States (1981) completed a study of U.S. industrial competitiveness with a focus on the steel, electronics, and automobile industry. This valuable report notes that the United States is currently engaged in a highly competitive global marketplace, but is unable to dominate as it did in the 1950's.

In absolute terms, much of American industry remains efficient and innovative, although in relative terms it may have declined with respect to other countries. For example, in the case of world steel production, the United States produced 26% of the world's steel in 1960 in contrast to 14% in 1980, while Japan increased its share from 6.4% in 1960 to 15.5% in 1980. While the United States retains technological superiority in many industries, it no longer holds any across-the-board advantage in the matter of technology. In the steel and automobile industries, which are currently facing stagnant or slowly growing markets, the United States is faced with choosing between maintaining competitiveness throughout the world market at the sacrifice of employment opportunities, or maintaining employment at the sacrifice of competitiveness. Nevertheless, a straightforward analysis leads one to realize that as competitiveness is allowed to erode, eventually U.S. firms will be unable to compete in any market and will lose most, if not all, of their employment due to a declining market share.

Several American industries have, in fact, maintained increasing productivity in contrast to other industries, as shown in Figure 1-3. U.S. competitiveness varies markedly across and within the diverse segments of industry. In the electronics industry it is greatest in high-technology sectors such as semiconductors and computers, though in international terms the U.S. consumer electronics industry is still rather small. For example, in 1978 almost 100% of the radios purchased in the United States were manufactured abroad, either by foreign firms or U.S. firms with foreign manufacturing plants. Approximately one-half of all phonographs and compact stereo systems were manufactured abroad. Although the United States has experienced extreme competitiveness in the consumer electronics sector, it has attained significant

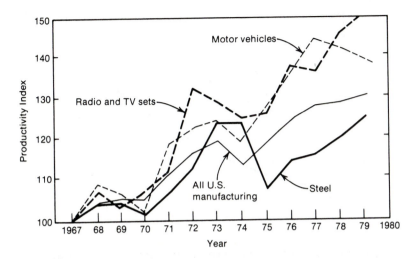

FIGURE 1-3.   Bureau of Labor Statistics productivity indexes (physical output per hour—all employees, 1967 = 100). *Source: Productivity Indexes for Selected Industries (Washington DC: Bureau of Labor Statistics.*

productivity increases in the areas of integrated circuit technology and has maintained a competitive stance with other nations such as Japan and the nations of western Europe.

International Productivity

Gross domestic production per employed person is shown in Figure 1-4. This figure shows that for a constant index of 100 for the United States, the nations of West Germany and Japan have been steadily gaining in their productivity measures, yet still have not attained the productivity maintained by the United States. Nevertheless, the United States, Europe, and Japan are becoming increasingly competitive and are locked in a world-market struggle.

As shown in Figure 1-5, U.S. productivity increased as the number of farm workers declined. Following 1970, the number of farm workers essentially leveled out and increases in the service sector caused reduced productivity gains. However, as shown in Table 1-2, some industries, such as the synthetic fiber industry, were able to maintain a productivity increase somewhat commensurate with the increase of hourly wages awarded to employees during the period 1974 to 1979.

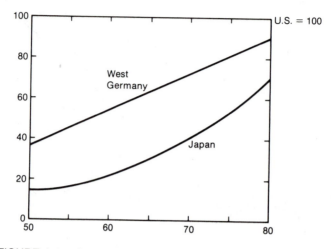

FIGURE 1-4.   Gross domestic production per employed person.

In contrast, productivity in the coal mining industry actually declined while hourly wages increased 10% on an average annual basis. Similarly, the steel industry experienced wide disparity between productivity gains and hourly wage increases. Such significant differences between productivity and labor cost result in inflated prices and re-

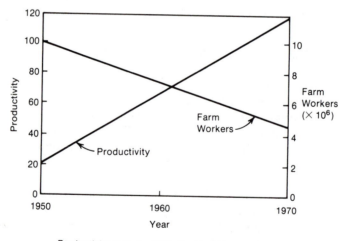

Productivity = Output/Worker-Hr (1967 = 100)

FIGURE 1-5.   U.S. productivity increased as the number of farm workers declined.

TABLE 1-2. Average Annual Percent of Change of Productivity
and Hourly Wages, 1974–79

Industry	Productivity	Hourly Wages
Synthetic Fibers	7.8	9.5
Telephone Communications	7.2	10.5
Tires	4.5	9.5
Steel	1.5	10.5
Coal Mining	−2.5	10.5

duced international competitiveness. The U.S. is currently inundated with foreign goods, and the imports do not stop with cars and steel: According to the Commerce Department, eight out of every ten metal nuts used in the U.S. now come from overseas (Burck, 1982). Many American steel mills purchase new steel mill equipment from Europe and Japan. In a recent study of industrial competitiveness for 1981 (Ball, 1982), the Geneva-based European Management Forum found Japan performing more than twice as well as Switzerland, the U.S., and West Germany, its closest rivals (see Table 1-3).

Figure 1-6 illustrates the rise in compensation per man-hour in the industrial sector for the period of 1952 to 1980. During this period the output per man-hour tended to decline while the labor cost per unit increased. As shown in the figure, the consumer price index increased relatively in concert with the unit labor cost. Of course, there are other factors that cause the consumer price index to increase, such as increased energy costs.

TABLE 1-3. International Industrial Competitiveness for 1981

Nation	Index
Japan	100
United States	48
West Germany	47
France	22
Britain	20
Italy	14

Source: R. Ball, Europe's Durable Unemployment Woes. Fortune, January 11, 1982, p. 70. From European Management Forum rankings based on 180 statistical criteria plus 60 subjective items. Among them: production costs, the adaptability of each economy to change, and "sociopolitical consensus."

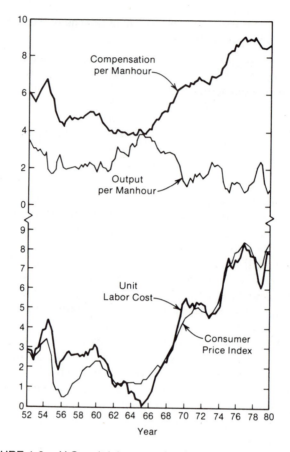

FIGURE 1-6. U.S. unit labor cost for the period 1952 to 1980.

The Automobile Industry

Automotive technology, like that of steel, is well diffused internationally and no one country has a technological advantage. Also, technical change in these industries is slow compared to electronics, and therefore major innovations are infrequent. However, the U.S. automobile industry is undergoing long-term internal restructuring and is experiencing significant short-term problems. The decrease in demand for automobiles since 1978 has combined with a shift in the market toward small cars and resulted in sharp declines in U.S. auto production and employment. The industry has experienced considerable financial losses in recent years.

While the United States automobile industry has maintained reasonable productivity increases, competitive nations have maintained equal productivity increases, and in some cases have actually outdistanced the U.S. For example, a recent analysis by James E. Harbour suggests that the average Japanese compact car is assembled in 14 worker hours against 33 for the comparable U.S. car (Burck, 1982). In other words, the average Japanese auto worker produces about 50 cars per year compared to about 25 for the U.S. worker. (A similar advantage is found in the steel industry, where the average Japanese steel worker produces 421 tons versus 250 tons in the United States.)

The Japanese also appear to have production cost advantages over the U.S.; in the subcompact market, the advantage amounts to 20% or more. The Japanese have achieved a total cost advantage of about $1,500 to $2,000 per car, of which approximately half is due to lower wages and fringe benefit costs in Japan; the Japanese auto industry wages currently run about $8.00 per hour less than in the U.S. The American practice of automatic wage gains linked to inflation rather than productivity is a major factor in the increase of the U.S. cost per automobile beyond that of Japan.

With Japan's factory automation methods and heavy capital investments in the automobile industry, the United States is finding itself at a disadvantage in the competition for the world automobile market.

The cost advantage of the American-made automobile can be illustrated by examining the changes in the cost of a Ford automobile over the period 1904 to 1982, as shown in Table 1-4. By means of innovation during the period 1904 to 1924, Ford was able to reduce the cost of its touring car significantly and the price-to-wage ratio was reduced by a factor of 18. In 1904, the average U.S. worker had to work 3.72 years to buy the 1904 Ford touring car; by 1924, he had only to work .2 of a year for a model of that year. By 1982, the cost of a comparable automobile had risen significantly and the price-to-wage ratio had increased again to one-third (.33).

TABLE 1-4. The Price to Annual Wage Ratio of an Average U.S. Worker for a 4-Cylinder Touring Car

	1904 Ford 2 Touring Car	1924 Ford Model T 4-Cylinder Touring Car	1982 Ford Escort 4-Cylinder
Cost	$ 2,000	$ 290	$ 7,000
Average Annual Wages	$ 538	$ 1,427	$ 21,000
Price/Wage Ratio	3.72	0.2	0.33

Factors of Productivity

Declining productivity remains the number one problem of American industry and much of the industry throughout Western Europe. The factors that contribute to improved productivity are illustrated in Figure 1-7. Technological innovation, increased automation, improved market share, improved worker satisfaction and motivation, beneficial government regulation and taxation, effective organizational structure, and improved maintenance of quality or raw materials are all factors that lead to increased productivity. When a firm allows the automation and capital equipment to age and deteriorate, it allows its productivity to decline. The United States has allowed its capital investment to deteriorate: 34% of the machines in U.S. industry are more than 20 years old, and over 69% of the cutting tools used for machining are more than ten years old.

The correlation between productivity growth and industrial capital investment for the period 1960 to 1973 is shown in Figure 1-8. This chart illustrates the potential for improving the productivity growth of a firm or industry by improving and increasing industrial capital investment. Japan has outdistanced Western Europe and the United States in the encouragement of industrial capital investments and continues to do so.

Productivity may be represented as influenced by three factors: labor, capital investment and technological innovation. This relationship can be represented in the equation

$$P = .14L + .27C + .59T$$

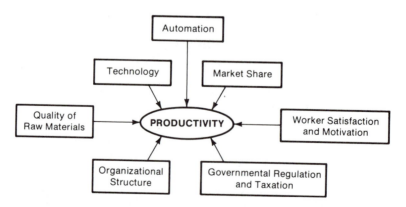

FIGURE 1-7.   Factors contributing to improved productivity.

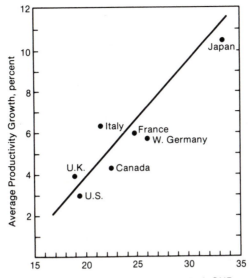

FIGURE 1-8.   Correlation between productivity growth and industrial capital investment (1960–1973). *Source: R.C. Deen, Jr., Technical Innovation, U.S.A.* Mechanical Engineering, *November 1978.*

where P represents productivity, L is labor, C is capital investment, and T is technological innovation. In this formulation, labor is represented as improved labor capabilities and reduced cost, capital investment as capital for automation in improved machines, and technological innovation as development of new devices and techniques for the improvement of the automated factory.

During the 1970's, research and development funding leveled out of decline as shown in Figure 1-9. Similarly, capital investment per man-hour of labor declined during this period. It is estimated that the investment rate in Japan is a third of their gross national product (GNP), while in the United States it is approximately 15% of the GNP. The problem in the United States is not simply declining innovation, but the slow adoption of innovation. Many innovations in automation developed in the United States are adopted more rapidly in Japan than they are in this country.

The United States national productivity program, then, should encourage an increase of diffusion of technical knowledge, an improvement of capital investment per man-hour of labor worked in industry or factories, and improved approaches to labor and quality of work.

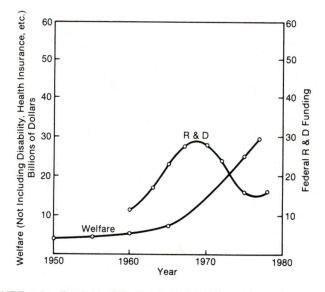

FIGURE 1-9. Federal welfare and R & D funding in constant 1972 dollars. *Source: Statistical Abstracts of the U.S.*

Attention must be paid to improved manufacturing methods and further education of workers and manufacturing engineers. Production has been neglected too long within the U.S., and the manufacturing function must again become one of the nation's prime considerations.

The benefits of improved productivity are many, as shown in Figure 1-10. As firms improve their productivity, they can hope for

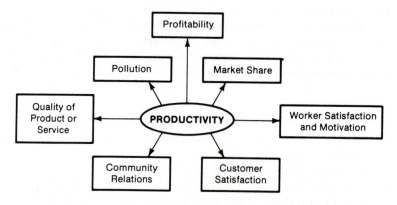

FIGURE 1-10. The benefits of improved productivity.

"So, by a vote of 8 to 2 we have decided to skip the Industrial Revolution completely, and go right into the electronic age."

FIGURE 1-11.   Cartoon by Sidney Harris. *Reprinted by permission.*

increased market share, improved worker satisfaction, improved consumer satisfaction and community relations, higher quality of product, reduced pollution and, of ultimate value to any firm, improved profitability.

The United States has contributed extensive technology in automation throughout its history. In 1805, Eli Whitney developed his method of interchangeable parts, which led to the development of the assembly line by Ford in 1905. Numerical control automation machines were developed in the United States in 1950. In the 1980's we can expect the emergence of automatic factories which will use automation in conjunction with improved manufacturing methods and improved labor methods to obtain improved quality and productivity. Many nations and peoples will strive to avail themselves of automation methods (see Figure 1-11).

2

Computer-aided Manufacturing and the Automated Factory

Since the onset of the industrial revolution in the late 18th century, industrial production has entered into a new phase. From individual, craftsman-like production, where each workman had to be skilled in all of the various aspects of his work, a form of production has come into being in which each workman need only be skilled in a fraction of the total work. The limits on production imposed by the capacity of an individual are thus overcome. This qualitative change effectively absorbed a large number of unskilled workers and led to the birth of modern large-scale industrial production. Since the dawn of the industrial revolution, advances in scientific knowledge, development of energy and other natural resources, and the pioneering work of Whitney, Ford, and others in the field of mechanization and mass production have caused industrial production to rise in unprecedented proportions.

Although manufacturing is thought to be a highly efficient activity, this is often not the case. Much effort has gone toward improving manufacturing processes by improving machine tools, materials, plant layouts, material handling equipment, and organization of the plant. However, little effort has been made to systematically incorporate manufacturing considerations at the product design stage so that difficult and costly manufacturing operations could be anticipated and eliminated by design modifications. Product design, which accounts for a major portion of the costs and difficulties of manufacturing assembly, holds large potential for cost reduction, since assembly and machining costs are a significant fraction of the total cost of the product.

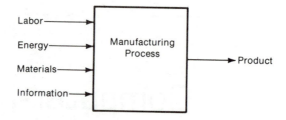

FIGURE 2-1. The manufacturing process.

The Manufacturing Process

The manufacturing process utilizes inputs of labor, energy, materials, and information in order to yield a product, as shown in Figure 2-1. It is the efficient combination of the input factors through a well-designed and controlled process that yields a high-quality, low-cost product. Over 35% of the work force of industrial nations throughout the world is engaged in the manufacture of piece goods as contrasted to continuous flow commodities such as oil or chemicals. During the last decade, the cost of computing power continues to decline while the unit-labor

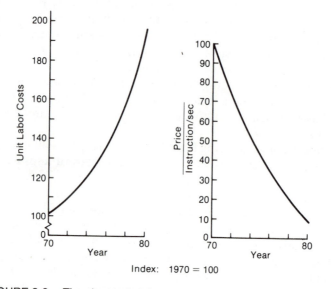

FIGURE 2-2. The change in labor costs and computing costs for the period 1970–1980.

FIGURE 2-3. (a) The activities required for developing and marketing a new product; (b) the manufacturing process.

costs have continued to rise, as shown in Figure 2-2. We can expect a shift in the manufacturing process toward an increasing use of information processed through computers as a substitute for labor, energy, and materials (Hudson, 1982).

The activities required for developing and marketing a new product are shown in Figure 2-3a. The process proceeds through research to development, to design, and then to production and marketing. The production or manufacturing process includes a number of steps, as shown Figure 2-3b. With the use of computers, the trend is moving rapidly toward highly automated factories utilizing computers in coordination with robots and other manufacturing equipment.

Numerical Control Machine Tools

According to many manufacturing experts, the use of numerical control (NC) machine tools has started a second industrial revolution. NC technology permits complex parts to be fabricated rapidly and accurately by automated machine tools that drill, grind, cut, punch, and mill to turn raw material into finished parts. NC machine tools were developed in the 1950's to fabricate contoured aircraft parts. Since then, the technology has spread throughout industry with machine tools and control systems.

NC machine tools use a numerical control method for activating the tools in response to a predetermined command stored as digital data on punch tapes, magnetic tapes, or in semiconductor memories. They perform two functions: positioning the tool point in three dimensions relative to a work piece, and controlling secondary functions such as speed, feeding, coolant flow, gauging, and tool selection. Figure 2-4 contains a block diagram of an NC machine tool.

Advantages of NC machine tools are:

1. One person can operate more than two NC machine tools.

2. NC machine tools have higher accuracy with better repeatability over conventional standard machine tools.

3. NC machine tools can process products which have complex theoretical shapes.

4. Shorter processing time can be attained by NC machine tools.

There are basically three types of NC machine tools. The first is the conventional NC system where functions are wired together in a fixed, preengineered arrangement. The second is a computer numerical control (CNC) in which a minicomputer is used to perform some or all of the basic NC functions in accordance with control programs stored in the computer's memory. Since the late 1960's, CNC has been gaining a significant share of NC machine tool industry. The third approach is direct numerical control (DNC), in which a group of NC (or CNC) machine tools are simultaneously controlled by a host computer. An example of a powerful CNC machine is shown in Figure 2-5.

Manual programming was the first method of preparing NC instructions. In this approach, the program is prepared from an engi-

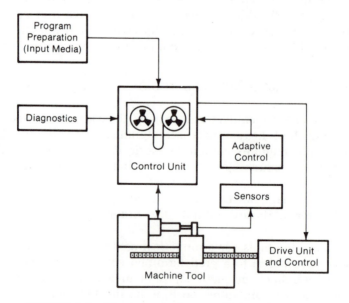

FIGURE 2-4.   The numerically controlled machine tool.

FIGURE 2-5.   The Bridgeport Series I CNC with automatic tool changer. The machine provides three-axis contouring with repeatability of
± .0005 inch. The single arm 24 position tool changer accepts tools
up to 4″ diameter and 10 lb. weight. *Courtesy of Bridgeport Machines
Division of Textron Inc.*

neering drawing in an alpha-numeric format suitable for direct entry
into the machine control unit. With computer-assisted systems, programming languages allow the user to develop NC instructions with
special commands, and computer processing routines perform com-
putations and put instructions in the proper format for specific machines.

A variety of specialized part-programming languages are available
for particular types of machine tools. In a step beyond this level of NC
programming, a new generation of software has emerged that uses a
geometric model created in computer-aided design as a basis for producing NC instructions. Thus, it is the combination of the computer
with its design and programming capabilities in coordination with the

numerically controlled machine tool that permits rapid, accurate, and complete manufacturing of products.

Computer-Aided Design

With the advent of low-cost computer capabilities, computer-aided design has emerged as a flexible method for design of parts and equipment. *Computer-aided design (CAD) is defined as designing with graphics using the capabilities of a computer to create, transform, and display pictorial and symbolic data.* As such, CAD helps organize and transform masses of often unintelligible raw data into the meaningful information essential for manufacturing. CAD, coupled with a capability of NC part manufacturing, is specifically termed *computer-aided manufacturing (CAM).*

With the evolution of NC machines, manufacturing resource planning, and computer-aided design, the computer has led to an integrated approach we now call CAD/CAM, as shown in Figure 2-6.

Traditionally, design involves transforming a concept into design information that is communicated to others through a drawing and related documents. However, this method is cumbersome, expensive, and requires a large number of designers; also, there is no convenient way to interact with the information in response to a demand placed on manufacturing. CAD/CAM provides an opportunity for productivity improvement in the design and manufacturing workplace. An example of a CAD/CAM system is shown in Figure 2-7.

In 1980 machine tool shipments amounted to $4.7 billion while

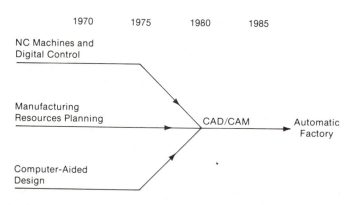

FIGURE 2-6. The evolution of automation in the factory leading to CAD/CAM and the automated factory.

FIGURE 2-7.   The Computervision Designer System is a CAD/CAM used to create, display, analyze and store engineering designs, to generate engineering drawings and reports, and to format design descriptions into information used to run automated manufacturing machines. The work station in the foreground is the primary input device. The processor and magnetic tape drive are shown in the right background, and the disk storage device is shown in the left background. The work station includes a cathode ray tube, a light pen input, a key input, and a printer output. *Courtesy of Computervision Corporation.*

CAD/CAM system sales amounted to $500 million. It is estimated that the CAD/CAM market is growing at 35% a year and will amount to $2.5 billion in 1985. The CAD portion of the system results from the combination of a computer with associated graphics and data base. This computer enables the designer to work with an evolving design, complete the drafting, and analyze the interim designs. Following the completion of a computer-aided design, the results are directly transferred into the system via computer programs, resulting in the computer control of the production machines by means of NC machines in combination with programmable controllers and robots.

Advances in the two primary elements of factory computer systems, computer-aided design and computer-aided manufacturing, will create a new industrial revolution. By integrating design with manufacturing, industrial firms are able to not only turn out a new product design much faster, but also program the computer to make sure the design provides quality and reliability as well as the lowest possible manufacturing costs. It is a common data base of part and product geometry and related information which makes it easier to translate a creative idea into a final product at a reduced cost. With CAD, a user can define a part shape, analyze stresses and other factors, check mechanical actions, and automatically produce engineering drawings from a graphics terminal. When CAD is combined with the CAM system, the user can manipulate nongraphic data such as bills of materials, shop information, and cost factors. The end result is greater design flexibility and what is referred to as "designing to cost." The CAD functions can be grouped into four categories:

- design and geometric modeling
- engineering analysis
- kinematics
- drafting

The number of CAD/CAM installations is growing by more than 30% per year (Krouse, 1982). CAD/CAM results in increased productivity, and CAD/CAM systems typically pay for themselves in less than two years. In addition, designs are developed closer to an optimum through the use of rigorous analytical techniques such as the finite element method. Cleaner, more accurate drawings are produced with high-speed plotters, and documentation is standardized with symbols and other parts of drawings stored in the computer and retrieved when needed.

One of the most recent advancements in CAD/CAM software is the development and continued refinement of solid modeling programs. This geometric modeling approach allows objects to be represented mathematically in the computer as solid forms. Developers ultimately hope to integrate all CAD/CAM functions into a unified system with a common data base. This level of sophistication will lead the way to the automated factory. The software available on many systems uses the geometric model created in CAD as a basis for producing NC instructions, so the user need not manually enter geometric data. Furthermore, machining information is provided through interactive question-and-answer prompts on the terminal. As a result, these systems allow NC instructions to be created graphically without requiring a knowledge of programming languages.

Eventually, as CAD/CAM is combined with robots and other automatic machine equipment in the factory, we will evolve towards the automatic factory.

Factory Automation

With the resurgence of interest in manufacturing technology, the United States is increasing its investment in putting the new automation technology to work. Some executives predict that by 1990 the U.S. will surpass Japan and Germany in the race to automate and will thus maintain the U.S. lead in industrial productivity ("The Speedup in Automation," 1981). The transformation is rooted in the realization that in two absolutely vital areas—computer software and computer-aided design—no other country comes close to matching U.S. know-how. With this edge, U.S. companies are moving towards manufacturing enterprises where computer control results in largely integrative computerized facilities. Purchases of computer-aided design systems have been climbing 35% or more annually, and should result in an estimated $2.5 billion market in 1985.

In the U.S. today, an estimated 75 percent of capital goods are produced in batches of 1000 pieces or less. CAD/CAM promises to bring automated batch-production of goods in runs of less than 50 units. Thus, while automation was formerly used for large throughput systems with mass production, we can now automate plants and change models and batches and still attain effective productivity through CAD/CAM. The cost of manufacturing one unit for a range of production quantity is shown in Figure 2-8. It is estimated that a CAD/CAM system can cut manufacturing lead time 25% and increase productivity by as much as four times the current level ("The Speedup in Automation," 1981). With the advent of the microcomputer, the computer intelligence can be brought to the individual machine and yet enable the intercommunication between microcomputers and overall supervisory computers and ultimately to the CAD/CAM date base. Analyses from certain early users of the new systems credit substantial productivity gains to CAD/CAM installations, due to increased machine tooling utilization and decreased worker hours.

The development of software and programs will be the key factor in expanding applications for CAD/CAM systems. Methods need to be developed for standardizing descriptions of shapes for the design phase,

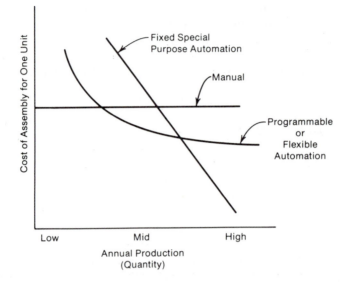

FIGURE 2-8. The cost of assembly or manufacturing one unit for a range of production quantity. Fixed automation, often called hard automation, is fixed or dedicated to one particular task throughout its life.

and for CAD systems, CAM computers, and the general purpose computer used by corporate headquarters to communicate with each other and share the information in their data bases.

The ability to make tooling and equipment changes quickly, easily, and at minimum cost in assembling different product models is becoming increasingly important to manufacturing managers. The need for flexible assembly and manufacturing operations leads to automated batch assembly and manufacturing facilities. Batch assembly is costly, in that it requires producing and inventorying relatively large quantities and assemblies. In addition labor is expensive, and any idle time on the part of assembly operators while they await tooling and part changeovers can be intolerably expensive. Finally, production requirements as dictated by customer needs and/or sales forecasts are often best met by scheduling so as to minimize work in process time. This usually means mixed model assembly operations on final assembly lines. Thus, with the use of microcomputers and programmable controllers, industry is able to build flexibility into automated assembly systems.

An additional new manufacturing management technique that is used is called *group technology*, a manufacturing philosophy that takes advantage of the similarities among parts and the processes that are

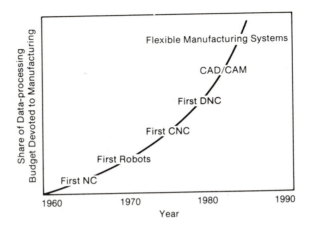

FIGURE 2-9.   The increasing cost of implementing computer-aided automation systems in the manufacturing plant.

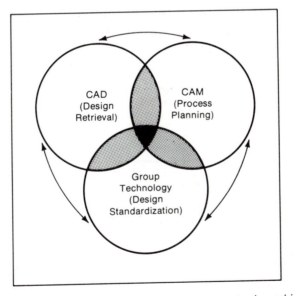

FIGURE 2-10.   Three key interrelated and computer-based ingredients of the automated factory are computer-aided design, computer-aided manufacturing, and group technology. CAD involves product design, CAM, product fabrication, and GT, product classification and coding.

used to manufacture them. Instead of treating each part as unique, parts are grouped into families, and codes are developed to identify parts characteristics such as size, shape, material, and manufacturing processes used. Because group technology classifies parts into groups with similar shapes or manufacturing processes, the parts can be manufactured in these families rather than individually. This reduces down time for tooling and set-up, and brings the advantage of long parts runs to operations producing relatively small lots. Group technology can also reduce machine set-up time by reducing the scope of tooling changes.

With the increasing use of computer-aided automation, a greater share of the data processing budget is allocated to manufacturing, as shown in Figure 2-9. The three energy-related and computer-based ingredients of the automated factory are computer-aided design, computer-aided manufacturing, and group technology, as shown in Figure 2-10. The increased use of these three approaches will yield improved productivity.

A computer material control system using material requirement

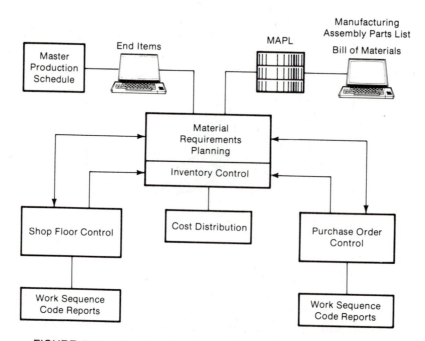

FIGURE 2-11. A computerized material control system using materials requirement planning (MRP). The MRP is operated using a computer and provides schedules and priorities for purchases, inventory, and shop flow control.

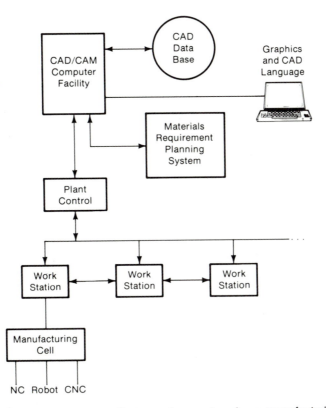

FIGURE 2-12. An overall automation system for a manufacturing process.

planning is shown in Figure 2-11. This system is also integrated with CAD/CAM.

One view of an overall automation system for a manufacturing process is shown in Figure 2-12. This system includes a CAD/CAM facility as well as a material requirement planning system, all communicating with work stations which control manufacturing cells consisting of NC machines, robots, and CNC machines. An example of a flexible manufacturing system is shown in Figure 2-13.

The Automated Factory

If a computer can control the machine directly, this leads to the natural evolution of computer-controlled multiple machines and eventually computer control of the entire factory. The totally automated factory is

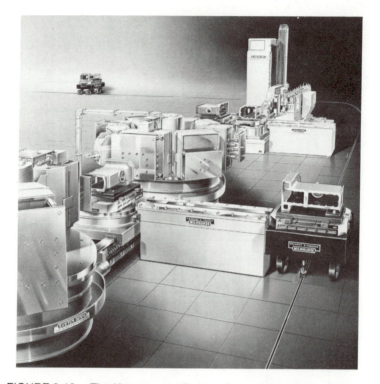

FIGURE 2-13. The Kearney and Trecker System Gemini is a flexible manufacturing system for mid-volume and mid-variety parts production. It incorporates an NC machine tool, a multiple-spindle head NC machine, and other special machine tools. The system incorporates a transporter using a palletized fixture and work piece holder. *Courtesy of Kearney and Trecker Corporation.*

a manufacturing facility which processes raw materials or components into finished products without direct human intervention. *Factory automation can therefore be defined as a process without direct human activity within the process.* The human would only be involved in designing the system and monitoring its operation. Realistically, this objective is decades away. In the near term we can look for more automated material handling, intelligent work stations for processing fabrication and assembly, and for integrated automated inspection functions as well as computer-aided product design. The worker's involvement in an automated factory would be in a control center from which all operation could be remotely monitored, or in equipment maintenance, or in computer-aided design of the original product. All of the industrialized nations are working toward the development of automated factories and most countries have examples of such systems in operation.

Perhaps the most impressive automated system is the assembly line at Seiko, Japan, which has developed a totally automatic system for the assembly of watches—no direct human input is used. In the United States, one example of an automated factory is that of Walter Kidde, Inc., a manufacturer of fire extinguishers in Belleville, New Jersey. This manufacturing operation is fully automated in that it starts with a billet of steel which is successfully drawn and shaped, heat treated, machined, assembled, tested, painted, and packaged, and comes off the end of the line as a finished product, untouched by anyone during the process. Another example of an automated factory is the West Coast facility of General Motors which automatically produces a complete automobile wheel—it enters at one end as strip steel and coiled bar, is automatically rolled, cut, formed, welded, punched, assembled, painted, and dried, and emerges at the opposite end as a completely finished product (Dallas, 1980).

The German firm of Messerschmitt-Bölkow-Blohm GmbH (MBB) has an aircraft manufacturing plant at Augsburg, Germany. Constructed between 1975 and 1978, the factory is totally automated including materials, requirements planning, and NC machines, and utilizes robots and automatic transporter systems. This plant is engaged in computer-aided manufacturing, since the design is done elsewhere. A unique tooling assembly and storage and delivery system has enabled MBB to increase its machine utilization from 50% available time to 75% (Link, 1981).

Computer-Aided Manufacturing*

The use of computers to control manufacturing has long been routine in continuous-process industries, such as chemicals and oil refining, but only in the past several years has it been introduced in batch-manufacturing industries, such as metal working. The metal-working industries employ by far the largest share of industrial workers in all the advanced countries, and a large percentage of their products are manufactured in medium-sized batches, ranging from hundreds to thousands of products in each batch. It is not feasible to produce such batches by rigid mechanized techniques, like those used in the large-batch automotive industry. To speed medium-range production, the flexibility of computerized automation is needed.

*From IEEE Spectrum, November 1981, pp. 34–39. Reprinted by permission.

It now appears likely that the main technical barriers to automation of most medium-batch manufacturing will be overcome in this decade. Rapid progress in the development of computer-aided manufacturing (CAM) in all industrialized countries will almost certainly lead to the construction of a prototype fully automated factory sometime in the mid-to-late 1980's. However, despite this rapid progress, it remains unclear to what extent industries will apply this potential for automation, especially in the United States.

At the moment, serious economic obstacles have prevented widespread use of computers for industrial control in American factories, although they have long been standard equipment in every sizable office. Many experts are worried that the Japanese are establishing a wide lead in putting electronics to work in improving productivity.

Wide Productivity at Stake.   The problem is not that the Japanese have developed superior automation technology. In fact, in state-of-the-art technology, the U.S. is equal to Japan in most areas and is clearly ahead in some, such as software development. But at present Japan is moving far faster in getting automation from the laboratory to the shop floor.

The metal-working industries that produce medium-sized batches are likely candidates for such automation because they produce all machinery used to manufacture everything else. Increases in the productivity of this sector would therefore decrease the cost of capital goods and could have a profound impact on the productivity of all industry.

Computer-programmed automation of the metal-working industries would not only ease the shortage of skilled machinists; it would also increase the productivity and versatility of the machine tools themselves. In traditional, nonautomated metal working, each machine is actually cutting metal only about 5 percent of the time, whereas in automated systems metal-cutting time may approach 70 to 100 percent of the available time, leading to from tenfold to twentyfold increases in tool productivity. This gain is especially important as machine builders move toward higher and higher cutting speeds, thus increasing the amount of metal cut for each minute the tool is operating. But such gains in cutting speed are useless unless the cutting time itself is increased to a substantial fraction of the overall workday.

Commercialized technology now exists for the partial automation of this metal-working sector. The most advanced equipment is the *flexible machining system* (FMS), which contains programmable machine tools and transfer devices to take parts from one tool to another, all under central computer control. FMS was introduced in the early 1970's, and nearly 80 such systems are now operating in Japan, the U.S., the Soviet Union, and the rest of Europe. Developmental work toward full-scale automation is far advanced.

The three primary technical problems are to increase the versatility of the systems; to automate the now quite heavy burden of maintenance; and to automate the difficult area of assembly of machine components. A large, government-funded project in Japan is aimed at accomplishing these tasks and constructing a prototype unmanned plant by 1984, while parallel, more modest efforts are being pressed in the U.S. and other countries.

Japan in Forefront of Implementation.   But in all countries a large gap exists between what is technically possible and the general implementation of computer-driven automation—even in Japan, the most ambitious in implementing computer-controlled manufacturing technology. Nearly half of existing FMS units, and approximately the same fraction of the world's industrial robots, are in Japan, which manufactures virtually all the robots and FMS systems used there. Japan is also the leading user of less sophisticated numerically controlled machine tools. The United States, Western Europe, and Eastern Europe including the Soviet Union, although close to the Japanese in development, lag considerably behind in implementation. Sophisticated automation remains a rarity in the U.S. and, to an even greater extent, in Western European metal-working industries. The Soviet Union and the rest of Eastern Europe have pioneered the use of extremely large-scale FMS technology, but such use has not become as widespread as in Japan.

In all countries the primary economic obstacles to the use of existing computer-controlled manufacturing technology have been the large initial capital investments needed and the relative uncertainty of the world market for machinery—a problem that has begun to affect even Socialist-bloc machine builders, whose production is at least partly oriented to a Western export market.

The State of the Art.   A few examples can illustrate the state of automation achieved by present commercial hardware. In the U.S. a typical system is the Kearney & Trecker FMS, installed in half a dozen factories. These systems contain between 5 and 12 machine tools, including complex machines called head changers. The latter can shift the heads on cutting tools to perform different tasks, such as drilling, boring, and milling. Parts, loaded onto pallets, are moved from one machine tool to another through use of an automatic towline system.

The FMS is controlled by a three-level hierarchy of computers. Overall control is provided by a mastercontrol minicomputer, which coordinates system production as a whole. The master control monitors the system for any breakdown of tools, machinery, or transports and alerts the supervisors to any breakdown. It schedules the work of each machine and routes the parts to the appropriate machines to maximize machine use and production.

Subordinated to the master control is a second minicomputer, the direct numerical control (DNC) module, which supervises the operation of the machine tools themselves. The DNC computer selects the programs to be carried out by the machine tools, transmits them at the appropriate time to the tools, and keeps track of the completion of cutting programs for transmittal to the master control.

The bottom layer of computer control is provided by the computerized numerical control unit (CNC) attached to each machine tool. The CNC receives the program from the DNC controller and then executes it, directly controlling the actions of the machine tool. The CNC also contains diagnostic programs that can detect mechanical or electronic malfunctions in the machine tool and report them to the central controllers.

Family of Parts Produced.   A typical FMS produces a family of parts, such as crankcases. Because of the much higher proportion of machine time spent cutting metal, the 15 stations of this system can replace about 90 stand-alone tools—a great saving in capital.

Other designs of such systems, such as the Cincinnati Milacron Variable Mission Manufacturing System, have improved in flexibility and can handle more than a given family of parts. Such newer systems have more fully automatic tool-handling procedures, to equip each machine station with its full complement of cutting tools. They also have greater random-access capabilities, to adjust the order in which parts move from station to station. Instead of fixed towline transports, the systems have wheeled, computerized vehicles that can follow a branching grid of wires embedded in the shop floor.

Similar technology is represented in Western Europe's most advanced FMS, the CIAM system at the Messerschmitt-Bölkow-Blohm plant in West Germany. This system, which manufactures wing assemblies for fighter planes, contains 25 multiple-spindle machine tools linked by a random-access mobile pallet-transport system. Tool changing and delivery are completely automated under central computer control, although the setting up of large pieces at the machines is partly manual. Because of the relatively long machining times needed to fashion the titanium wing parts, the operator can set up one part while another is being machined; thus the tools are continuously occupied.

The East Germans have developed computerized machining systems for extremely heavy parts of up to 24 tonnes, and they have begun to standardize the modules that make up large automated machining systems. In the ROTA FZ2000 system in Erfurt, for example, the 52 work stations in the system are constructed from standardized modules. The system turns out a million gears a year in small batches.

The Soviet Union has also emphasized standardization in developing FMS for a variety of industrial uses. The ASKI system has standardized tools, spindles, transfer mechanisms, CNC controls, and sensors, all under the control of a single computer.

Automating the Night Shift.   In Japan many FMS units are aimed at the operation of unmanned third—or night—shifts in the machine-tool industry. Very few machinists in any country are now willing to work night shifts, particularly in the U.S., where industry has trouble finding enough machinists to fill two full shifts.

A new Fanuc company plant in Fuji, producing electrochemical spark-erosion machine tools and robots, is an example of the solution to this problem. The plant has 29 automated work stations connected by unmanned vehicles that are guided by optical methods. Robots at seven of the stations load and unload workpieces. Materials and finished products are stored in two automated warehouses. During the day shifts, 19 workers are in the machining section, making up the pallets that carry the parts, and 63 are in the nonautomated assembly section. At night the assembly section stops work, and the machining section continues operating with the pallets made up during the day. Only a single worker is needed to monitor operations via TV.

The Japanese have also adapted their FMS to what they call the "just in time" system of inventory control developed in the Japanese auto industry. In this system, the exact number of parts needed in later stages are made up in the early stages of a manufacturing process, minimizing in-process inventory.

In all of these examples, the machining process is the only one automated. Computer-controlled assembly is, in general, at a much less advanced stage of development and is not commercially available anywhere. Highly repetitive, mass-production assembly processes, such as in the watch industry, have long been mechanized, but they do not require the flexibility and programmability needed for most assembly tasks in the machinery industries, where optical and tactile sensors are needed.

Westinghouse, with the support of the National Science Foundation in the U.S., has developed a prototype adaptable programmable assembly system (APAS), which assembles end-bell complexes for electric motors. The APAS consists of two stations. The first uses a modification of a Stanford Research Institute computer vision system to locate the end-bell on a conveyor belt. The vision module then instructs a Puma robot arm to pick up the bell, orient it, and put it on pins for the next step. The second station takes the bell, adds various washers and attachments, and sends the assembly to a hydraulic press for completion. The APAS is admittedly only a small step toward the sophisticated assembly machines required by machine builders.

Toward Unmanned Factories. A number of technical problems must be solved in order to move from the present state of the art to fully computer-controlled, unmanned factories. Electronic hardware is not one of them. Microprocessor and minicomputer capacities are more than adequate today to control the most complex factory organization. Software problems are more critical, but they are yielding rapidly to developmental efforts in the U.S., probably the leading country in this area of computer-aided manufacturing. The Air Force-sponsored ICAM (integrated computer-aided manufacturing) project is producing a number of generally useful programs for a variety of control tasks and simulations.

ICAM is concentrating on the development of CAM for the aerospace industry, in particular for sheet-metal operations and composite fabrication and assembly. Though the project's main emphasis is on software and computer technology, some effort is going into hardware design and the development of computer-controlled manufacturing cells.

The main problems of automation in CAM are in the mechanical hardware itself and in the interfaces between the electronics and mechanical parts such as the visual sensory systems. A big area of concern is simply that of ensuring the mechanical reliability of the many moving parts of the system—an absolute necessity to avoid frequent and very expensive shutdowns of whole systems. This sort of problem plagued the early and highly ambitious East German efforts in automated assembly, maintenance, and repair functions. In assembly the crucial missing element is a computer vision system sufficiently sophisticated to deal with objects in three dimensions [see "Computers that see," *Spectrum*, October 1980, p. 28]. Hitachi has developed an experimental device with eight TV sensors and two arms for assembly work, and the company has used the techniques developed with this device to produce a programmable assembly robot.

The requirements for maintenance and repair robots are similarly involved, and again the Japanese are leading in meeting the requirements. Such robots are important because automatic systems require frequent repairs, and only with maintenance robots can the last worker be eliminated from night shifts. A number of Japanese universities have developed pattern-recognition systems to detect tool breakage either acoustically or through analysis of the electric current that is fed to the spindle motors. A national project has been initiated to develop mobile robots capable of independent repair operations. Although the robots in the project are intended initially for use in nuclear reactors, they will also be applied later to tasks in unmanned manufacturing.

Mobile robot development is also being pursued at the LAAS organization in France, where laser range-finding is being used to help the robot find its way about, and at Warwick University in Britain.

Better Machine Tooling Needed.   It will also be necessary to expand the capabilities of flexible manufacturing systems to produce a greater variety of machine parts. The main limitation here is that each family of machine parts is manufactured by a distinct mix of machine tools. Thus one type of part might require a lathe, a milling machine, and a grinding machine, and another might need a boring machine and a drill press. Though machining centers and head changers can perform a number of such different tasks on one machine, complete flexibility to perform any task on each machine would be highly desirable.

The Japanese government, as part of a national project for an unmanned factory, has been seeking a solution with so-called metamorphic machines. These consist of a variety of modules that can assemble themselves on command into any type of machine tools required for a given part.

The most serious obstacle Japanese researchers have found in designing such a remarkable machine is in achieving the essential rigidity for precision machining. Here computer-aided design techniques have been used, both to assist in the design of the modules themselves and to allow the controlling computer to check the rigidity of the assembled machines continuously during production.

Unmanned Factory Sought by 1984.   With its national projects, the Japanese government is attempting to combine all the steps needed for unmanned factory operation. The project, begun in 1977 by the Agency of Industrial Science and Technology of the Ministry for International Trade and Industry, aims at the completion of a fully automated factory in 1984. The factory is to produce modules of its own metamorphic machine tools. It will thus be a step toward a self-replicating machine.

The plant will consist of a small number of metamorphic machines called complex production systems, including a cutting station, a machining station, an assembly station, a laser processing station, and an inspection station. High-powered lasers will be used in cutting, welding, and heat-treating operations.

Although the Japanese originally planned to automate the initial rough-casting and metal-forming operations, they have found this too difficult at present. Instead these operations will be performed off line by conventional methods and the castings will then be fed into the plant. However, after that point—and up to the final inspection—the entire operation will be under computer control.

Another change in the conception of the project was to limit production to machine-tool modules instead of trying to produce any machine parts. As one of the Japanese researchers remarked to an American visitor, "We found that a machine designed to make everything would really produce nothing."

When completed, the fully automated plant is to produce components weighing up to 500 kilograms and measuring 1 meter in any dimension. About 10 technicians are expected to be needed to supervise production, but the plant is to have an output similar to that of a conventional factory employing 700 workers.

According to Eugene Merchant, principal scientist at Cincinnati Milacron in Ohio who recently visited the Japanese project center, the work appears to be on schedule, although some American experts have expressed doubt that the project will be completed on time. Prototypes of the metamorphic machines are being tested, and construction of the plant itself is expected to begin soon. Initial operation, after shakedown, is planned for 1983 or 1984. At present the Japanese are not making public any details of their work or allowing visitors to see the machinery, but if they have indeed built prototypes of all the machines required by the plant, they have established a long lead over automation efforts in the U.S. and elsewhere.

The impact of an operational automated machine-tool factory in Japan would be immense. Such factories would be able to produce duplicates of their own equipment quite quickly, and that would lower the cost of machine-tool production within a relatively few years. Not only would this give a further boost to Japan's already booming machine-tool exports, but it would also lead to rapid reduction of the costs of Japanese machinery manufacture generally, posing serious competitive problems for U.S. and Western European machine builders. Machinery is West Germany's leading export, and it is the second largest export of the U.S. after food.

Competitive Edge for Japan.   Such an automation breakthrough would widen the already significant lead that Japan has in putting computers to work in manufacturing. Japan has 30 FMS units at work in its industries, while the U.S., Western Europe, and Eastern Europe, including the Soviet Union, have only about a dozen systems apiece, despite the fact that each of these sectors has a considerably larger economy than that of Japan (more than double in the case of the U.S.). It should be noted that, as is the case with robot estimates, such numbers vary according to the definition used.

Japan has a similar lead in the use of less sophisticated types of computer-controlled automation. Of 13,000 robots installed in the capitalist economies, Japan with 6000 has nearly half, overshadowing the U.S., which has 3500, and Sweden and West Germany, with 1200 each. In addition, the U.S. and Western Europe import nearly 50 percent of their robots, while Japan, the leading exporter of robots, imports only 4 percent.

Japan is compounding its competitive advantage by using robots to produce robots, thus reducing their cost, as at the Fanuc plant in Fuji. Japan also leads the U.S. in its use of numerically controlled machine tools, the basic building blocks of automated machining systems. The majority of Japan's machine-tool production is now numerically controlled, while only a fraction of U.S. production is.

By contrast, the application of factory automation in the U.S. has proceeded slowly, although the first FMS units were installed at roughly the same time in both countries (1971–72). Since then only about one FMS a year has been installed in the U.S., chiefly in agricultural and construction machinery manufacture and in the aircraft industry. Although there has been a surge of interest in FMS's in the last year, this has not yet been translated into any spectacular rise in orders by manufacturers.

On the other hand, the U.S. robot industry is definitely beginning to take off. Considerably more sophisticated models have come onto the market in the last two years, increasing the range of applications. The relatively limited Unimate robot, from Unimation Inc. in Danbury, Conn., has been replaced as the dominant type by the Cincinnati Milacron T-3, a multiple-jointed arm that can imitate human elbow and wrist movements. The cost effectiveness of robots is impressive, with capital recovery through savings in wages in as little as 12 to 18 months. As a result, auto manufacturers in particular, along with machine builders, have been turning increasingly to robots to improve productivity. Robot sales in the U.S. last year were estimated at $200 million, and the production rate is growing nearly 30 percent a year. The leading robot application in the U.S. is welding, followed by machine loading and painting.

The level of application of computer-aided manufacturing in Eastern Europe, including the Soviet Union, is roughly comparable to that in the U.S. Development has been spurred by ever-worsening labor shortages, a problem especially acute in East Germany, where the population has ceased to grow. Many jobs, such as painting buses, are deemed too monotonous for people and are automated. In addition, Eastern European governments view automation as a way to enhance their position in the Western machinery export market, an important source of income to both East Germany and Czechoslovakia.

The importance that the Soviet leadership places on CAM is indicated by President Leonid Brezhnev's report to the 26th Congress of the Communist Party in 1981. He urged engineers in defense manufacturing to help disseminate computer manufacturing techniques to civil industries. Cooperative projects between Eastern European countries—East Germany and Czechoslovakia, for example—have sped the development efforts there.

At present Western Europe is lagging behind the other industrialized countries in CAM. Only West Germany has a major government-sponsored project, although Britain and France have recently initiated small projects. Leading Western European industries seem to be moving very cautiously, investing only in evolutionary developments that have the best prospects for short-term returns. Most companies are cloaking their moves in commercial secrecy.

Sweden is probably the most active of the Western European countries in CAM, because the virtual elimination of night-shift work in that country has put a high premium on the development of unmanned third shifts. By comparison, in Italy, which has a much larger economy, only about a third as many robots are used. They are concentrated almost exclusively in three companies: Fiat, Olivetti, and Alfa-Romeo. Robot use in Britain is even more limited.

As in the U.S., welding is the leading application for robots in Western Europe. In all the industrialized countries, the simpler applications of computer-controlled manufacturing, such as stand-alone robots, have been expanding much more rapidly than the bigger and much more complex FMS units.

U.S. Progress Hampered by Economics. The principal reason U.S. manufacturers are reluctant to move quickly with computer-controlled automation is the relatively large investment needed for each project and the uncertainty of markets. For the users of automated equipment, it is vitally important that systems, once purchased, remain in almost continuous use. However, the highly cyclical nature of machinery markets coupled with the general softness of such markets since 1975 tend to increase the risk that such steady demand will not be achievable.

The Carter Administration first set up and then, in 1980, abolished the Center for Productivity under the Department of Commerce. The center provided grants for university and industry projects in CAM. Much of this function was subsequently taken over by the Cooperative Technology Project, which, in turn, was abolished by the Reagan Administration in 1981.

The Japanese automation effort has not been hobbled by problems like these. Japanese industrialists tend to put more emphasis on long-term gains than do their American counterparts, and Japan's role in the world market encourages them. Since Japan has a smaller overall share of the world market than the U.S. does, Japanese manufacturers can invest in improved technology with the expectation that they will receive a larger market share, even if the world market shrinks. Japan's more rapidly growing economy is also less burdened with obsolete equipment than are U.S. industries, enabling investment to be channeled into improvements rather than just replacement equipment. This more rapid growth also gives Japan an extremely tight labor supply, which further encourages long-term automation decisions. Finally, the Japanese government is willing to take on risky industrial research and development, as in its national project, in sharp contrast with the reluctance of recent American administrations.

On the other hand, Japan's method of automation may be causing serious problems for its workers. Some Japanese trade unionists contend that highly automated systems have frequently been introduced without efforts to control their effect on working conditions. As a result, workers are often forced to work faster to keep up with the new machines. American trade unions, while generally not hostile to CAM, have insisted that automated machines not be used to speed up remaining manual tasks.

In spite of these problems, it appears that only a serious contraction of the world economy could blunt Japan's momentum in developing and applying CAM technology. It is much more uncertain whether the U.S. and Western Europe will be able to narrow Japan's lead. As Dr. Merchant of Cincinnati Milacron commented, "The Japanese have taken this technology and run with it, but the U.S. still hasn't."

The Future of Factory Automation

For decades, machine shops were the domain of skilled craftsmen. Today, the picture has changed as manufacturing has evolved toward the automated factory. Factories are becoming more flexible as well as more productive, resulting in improvements in both price and quality.

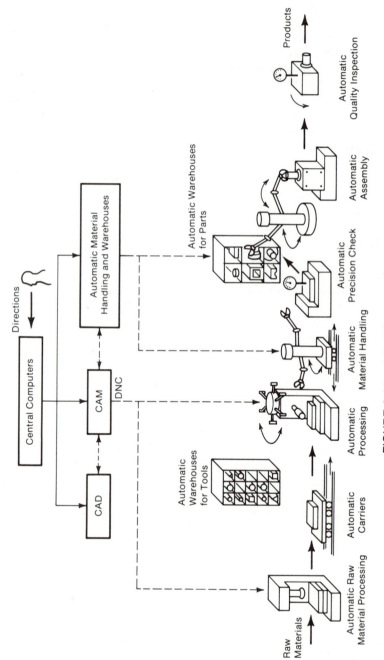

FIGURE 2-14. A flexible manufacturing system.

Central Computers

Directions

Automatic Material Handling and Warehouses

CAM

DNC

CAD

Automatic Warehouses for Parts

Automatic Quality Inspection

Products

Automatic Assembly

Automatic Precision Check

Automatic Material Handling

Automatic Processing

Automatic Warehouses for Tools

Automatic Carriers

Automatic Raw Material Processing

Raw Materials

Increased flexibility achieved through the use of computers will enable the manufacturing plants of tomorrow to incorporate robots, NC machine tools, computer-aided design, and computer-aided manufacturing systems, unmanned parts carriers, automatic warehouses, and sensors for control of the overall system. Thus the future plant will also be more controllable.

The flexible manufacturing systems of the future will combine the technologies of NC machine tools, computers, material handling systems, and industrial robots. They will also include computer-aided design systems and automatic warehouses. A diagram illustrating a flexible manufacturing system and its components is shown in Figure 2-14. Fujitsu Fanuc Company of Japan constructed a factory in 1980 costing 10 billion yen. It is the production center for a wide variety of electronic machine systems including industrial robots, CNC machines, and other machine tools. The nucleus of the Fujitsu system is a machining cell composed of CNC machine tools, an industrial robot, and

FIGURE 2-15. The flexible manufacturing system at the Murata Machinery plant in Japan uses a driverless vehicle that moves about the factory guided by wires implanted in the floor.

a monitor, and it utilizes unmanned carriers to move materials. An example of a driverless vehicle which is planted in the floor and moves about the factory guided by wires is shown in Figure 2-15. As the automated factory evolves, the need for improved computer software, programming capability, and control of machines and robots will be required.

3

The History, Development, and Classification of Robots

The concept of robotics, although not referred to by that term until relatively recently, has captured man's imagination for centuries. One of the first automatic animals—a wooden bird that could fly—was built by Plato's friend Archytas of Tareentum, who lived between 400 and 350 B.C. In the second century B.C., Hero of Alexandria described in his book, *De Automatis,* a mechanical theater with robot-like figures that danced and marched in temple ceremonies.

The precursors of programmable robots are classified as automata in contrast to toys because of their length and complexity of their operating cycles. Two examples of automata are shown in Figure 3-1.

A French engineer, Jacques de Vaucanson (1709–1782), was elected to the Academie des Sciences for his work, which included the creation of a life-size, flute-playing shepherd. Pierre and Henri Jacquet-Droz constructed life-like automata driven by springs (Spilhaus, 1982).

In 1921, Karl Capek, the Czech playwright, novelist and essayist wrote the satirical drama R.U.R. (Rossum's Universal Robots), which introduced the word "robot" into the English language. The playwright coined the word to mean forced labor; the machines in his play resembled people, but worked twice as hard. As shown in the photo of Capek's play in Figure 3-2, Capek pictured robots as machine-like human look-alikes, with arms and legs and personalities. The fact that this image still prevails today is illustrated by the character C3PO from the 1977 movie *Star Wars* shown in Figure 3-3, although the industrial robots in today's factories look nothing like humans. The Germans were the first to put robots on the screen, in a 1926 movie called *Metropolis.* In

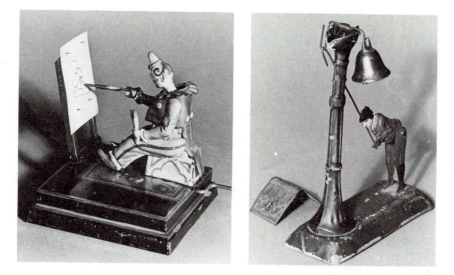

FIGURE 3-1.   The German clown (left), made in 1895, sketches a picture, his pencil controlled by a double cam in the base that can be changed to produce different pictures. The French bellringer (right). *Courtesy of Nelson McClary and A. Spilhaus.*

FIGURE 3-2.   In the play *R.U.R.*, the robots were manufactured for profit, as a replacement for human workers. The robots shown turned against their creators. Alquist, the clerk in the robot factory, is allowed the dubious honor of being the last human to die, because the robots regarded him as the worker who most resembled them. *Courtesy of New York Public Library Theater Collection.*

FIGURE 3-3.   The robots C3PO and R2D2 appeared in the films *Star Wars* and *The Empire Strikes Back* as the able assistants and supporters of Luke and Leia, the hero and heroine of the films. These robots were portrayed in anthropomorphic form. *Courtesy of Lucasfilm, Ltd.*

1939, Electro, a walking robot, and his dog Sparko were displayed at the New York World's Fair. In the same year science fiction writer Isaac Asimov started writing stories about robots. Asimov's stories fired the imagination of a Columbia University physics student named Joseph F. Engelberger. In 1956 Engelberger had a conversation with George C. Devol, the inventor of something he called a programmed articulated transfer device. By the time Devol's patent application was granted in 1961, Engelberger had started Unimation Inc., which bought the rights and built developmental versions of Devol's device, now called a robot.

In the early 1960's, George Devol and Joseph Engelberger introduced the first industrial robot through Unimation, Inc. (Froehlich,

1981). The idea was to build a machine that was flexible enough to do a variety of jobs automatically and could be easily taught or programmed so that if the part or process changed, the robot could adapt to its new job without expensive retooling, as was the case with hard automation. It was this mating of a computer to a flexible manipulator that has opened the door to new methods of manufacturing. Dr. James Ablus in a recent book on the effects of computers and robots, wrote:

> The human race is now poised on the brink of a new industrial revolution which will at least equal, if not far exceed, the first industrial revolution in its impact on mankind. The first industrial revolution was based on

TABLE 3-1.   A Chronology of Developments in Robotics

1770	Pierre and Henri Jacquet-Droz construct life-like automata that can write, draw, and play musical instruments and are controlled by cams and driven by springs.
1801	A programmable loom is designed by Joseph Jacquard in France.
1946	George Devol develops the magnetic controller which is a playback device.
	J. P. Eckert and John Mauchley build the ENIAC computer at the University of Pennsylvania.
	Bullard Company develops and sells the MAN-AU-TROL automatic machine control system.
1952	The first numerically controlled machine is built at MIT.
1954	George Devol develops the first programmable robot.
1962	General Motors installs its first robot from Unimation.
1968	An intelligent mobile robot, Shakey, is built at SRI.
1973	Richard Hohn develops T3, the first commercially available robot from Cincinnati Milacron.
1976	NASA's Viking 1 and 2 landers perform on Mars with their sample-collecting arms.
1978	The first PUMA arm is shipped to General Motors by Unimation.
1982	Unimation is acquired by Westinghouse.

the substitution of mechanical energy for muscle power. The next industrial revolution will be based on the substitution of electronic computers for the human brain in the controller machines and industrial processes. ("Automatic Factory: Identify Its Fingerprints," 1981.)

A chronology of developments in robotics is given in Table 3-1. The Man-Au-Trol of 1945 is shown in Figure 3-4, while the MOBOT developed is 1961 is shown in Figure 3-5. The Viking Lander, a complex space robot, is shown in Figure 3-6.

   The use of computers, sensors, and mechanical actuators or manipulators as a coordinated system utilized in manufacturing systems has been a subject of study and application for several decades. Manually controlled manipulators for space systems and for nuclear fuel control have been designed and implemented for over 25 years. Robots combine computer intelligence, modern sensors, and manipulator arms to provide flexible devices that can economically increase the productivity of manufacturing processes.

FIGURE 3-4.   The Bullard Company was cited for top honors for the development of the Man-Au-Trol Verticle Turret Lathe in 1945. This hydraulically actuated, electrically controlled system provided an automatic lathe. *Courtesy of The Bullard Company.*

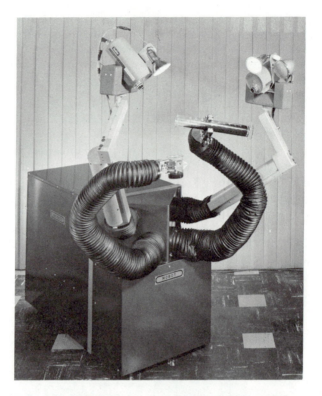

FIGURE 3-5.   The MOBOT developed by Hughes Aircraft Company. This multiple-arm robot using vision systems was developed as a prototype in 1961. To provide maneuverability, the wrist, elbow, and shoulder are double-jointed. *Courtesy of Hughes Aircraft Company.*

The robot manipulator task can be represented by the system shown in Figure 3-7. The sensors may include vision systems, touch feedback, parts identification and distance sensors among others. There are numerous designs of arms as well as of the computer intelligence systems. Within the next 10 to 15 years, with the increasing use of computers, we can expect further practical application of computer manufacturing and robot systems. Many computer factory systems exist today as islands of automation. We can expect that tomorrow more integrated factory systems will become economically viable. With the enhanced functions of computer software and hardware, the introduction of computers for industry automation becomes more prevalent.

FIGURE 3-6. The Viking Lander remote space vehicle is the most complex robot ever constructed, and has proved remarkably dependable. Here Viking I demonstrates its long boom arm for collecting soil samples. The Lander houses numerous experiments, all conducted under automatic control. Viking II landed on Mars on September 3, 1976. *Courtesy of Martin Marietta Corporation.*

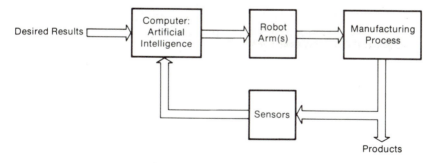

FIGURE 3-7. The robot task.

Definitions of Robots

The Electric Machinery Law of Japan defines an industrial robot as an all-purpose machine equipped with a memory device and a terminal device (for holding things), capable of rotation and of replacing human labor by automatic performance of movements. Japan classifies industrial robots by the method of input information and teaching as follows (Tanner, 1970):

1. Manual manipulator—a manipulator that is worked by an operator.

2. Fixed sequence robot—a manipulator which repetitively performs successive steps of a given operation according to a predetermined sequence, condition, and position, and whose set information cannot be easily changed.

3. Variable sequence robot—a manipulator which repetitively performs successive steps of a given operation according to a predetermined sequence, condition, and position, and whose set information can be easily changed.

4. Playback robot—a manipulator which can produce, from memory, operations originally executed under human control. A human operator initially operates the robot in order to input instructions. All the information relevant to the operations (sequence, conditions, and positions) is put in memory. When needed, this information is recalled (or played back, hence, the term "playback" robot) and the operations are repetitively and automatically executed from memory.

5. NC (numerical control) robot—a manipulator that can perform a given task according to the sequence, conditions and position, as commanded via numerical data. The software used for these robots includes punched tapes, cards, and digital switches. This robot has the same control mode as an NC machine.

6. Intelligent robot—this robot uses sensory perception (visual and/or tactile) to independently detect changes in the work environment or work condition and, by its own decisionmaking faculty, proceed with its operation accordingly (Engelberger, 1980).

In this book we use three different robot classifications:

1.   Robots by Japanese Definition—all 6 classes.
2.   Robots by U.S. Definition—classes 3, 4, 5, 6.
3.   Sophisticated Intelligent Robots—classes 5, 6.

The American Robot Industry Association (RIA) defines a robot as *a manipulator designed to move material, parts, tools, or specialized devices, through variable programmed motions for the performance of a variety of tasks.* Thus, the U.S. definition of robots eliminates the manual manipulators and fixed sequence machines (Krouse, 1981; Marsh, 1981). Another definition is *an automated articulated machine with feedback between the work space and control space.**

Robots are generally seen as anthropomorphic copies of humans, combining human qualities of intelligence, mobility, and manipulation ability. The robot's end effector or "hand"—it may be a spray painter, screwdriver, welder, or gripper—equates to a tool in the hand of a human. Depending on the intelligence that is provided within the computer controller and its computer, it can be a highly sophisticated robot fitting in classification 6. In a technical sense, industrial robots are universally applicable kinetic machines which, in contrast to conventional machines, can carry out relatively complicated series of movements.

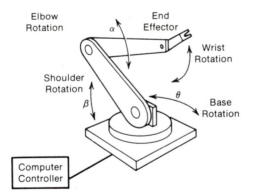

FIGURE 3-8.   Four-axis anthropomorphic motion, also known as revolute motion, is the most sophisticated. It can yield up to six degrees of motion in the shoulder, elbow, and wrist.

*Developed by Peter Bock of George Washington University and Gary Muhonen of Mountain Computer, Inc.

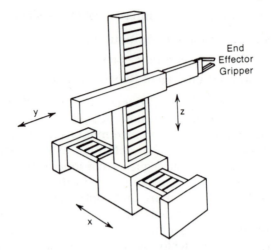

FIGURE 3-9.   Rectilinear or Cartesian motion. The robot's arm moves along three axes, *x, y,* and *z*.

The physical configuration of a typical robot is shown in Figure 3-8. This system illustrates a four-axis robot, which includes base, shoulder, elbow, and wrist rotation, as well as an end effector. A fifth axis can be added at the wrist to enable it to rotate left and right as well as up and down; a sixth axis is often added to enable the wrist to

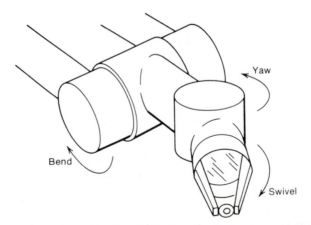

FIGURE 3-10.   The three-axis action of a wrist with an end effector. Bend is often called pitch, while swivel is often termed roll.

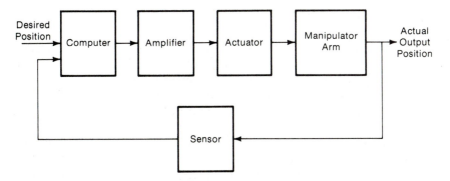

FIGURE 3-11.   The computer control of an intelligent robot.

swivel at the joint. A robot based on a linear axis motion, along a set
of linear axes, is shown in Figure 3-9. This system can be used effectively for tasks such as loading and unloading machines. The three-
axis action of a robot wrist with an end effector (gripper) is shown in
Figure 3-10. The intelligent robot system uses a computer to achieve
control of the manipulator arm as shown in Figure 3-11. By contrast,
the control of a simple, limited-sequence system is shown in Figure 3-
12. This system uses limit switches to signal completion of the desired
movement. Point-to-point motion is a type of robot motion in which a
limited number of points along a path of motion is specified by the
controller, and the robot moves from point to point rather than in a
continuous, smooth path.

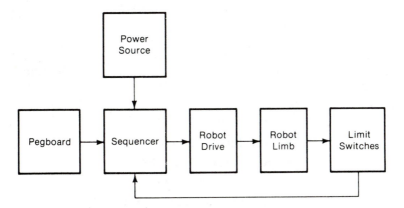

FIGURE 3-12.   A limited-sequence, point-to-point robot uses a peg-
board to arrange the desired sequence. The limit switches signal completion of the desired movement in one axis.

Robot Characteristics

Most industrial robots bear only partial resemblence to anthropomorphic robots. They are not mobile, and they have little more than
an arm attached to a stationary base. However, they all possess some
form of intelligence through their connection to a computer controller.
The controller functions as the brain and nervous system of the robot.
It can be a reprogrammable device ranging from a simple logic system
to a full digital computer. In sophisticated industrial robots, the control
computer is capable of a high level of artifical intelligence and not only
runs the robot through its program moves, but also integrates it with
ancillary machinery equipment and devices. The controller can use
other inputs and monitor processes and is able to make decisions based
on system demands while at the same time reporting to a supervisory
control computer.

The manipulator consists of the base and arm of the robot including the power supply which is hydraulic, electric, or pneumatic.
The manipulator is the component that provides movement in any
number of degrees of freedom. The hand or gripper, sometimes called
the end effector, can be a mechanical, vacuum, or magnetic device for
parts handling. At present, most of these devices are only two-position
and are either open or closed, i.e., on or off. The grippers are often
changed to accommodate different work tasks and possess little capability of their own. In the future, one can expect end effectors with
intelligence imbedded in the effector as well as a variety of capabilities
similar to those of the human hand and tools used in conjunction with
the human hand. However, the gripper of today's industrial robots is
one of the most limiting factors in universal robot utilization due to
the lack of hand programmability. This is the weakest link of the robot
components. Extensive research and development is being done to produce a gripper than can handle a wide assortment of part configurations
(Glorioso & Osorco, 1980). The Japanese use the word *mechatronics* to
signify the harmony of mechanical and electronic capabilities. An example of a current robot and gripper is shown in Figure 3-13.

The robot can be structured in four coordinate systems:

- Revolute (Figure 3-8)
- Rectilinear (Figure 3-9)
- Polar (Figure 3-14a)
- Cylindrical (Figure 3-14b)

The Unimate 2000 robot shown in Figure 3-15 operates in polar
coordinates.

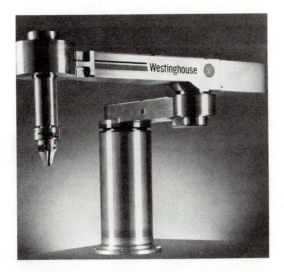

FIGURE 3-13.   The Westinghouse Series 1000 modular robot. *Courtesy of Westinghouse Corporation.*

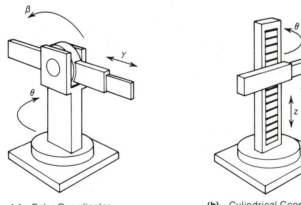

(a)   Polar Coordinates   **(b)**   Cylindrical Coordinates

FIGURE 3-14.   Polar coordinates (a) and cylindrical coordinates (b) for a robot arm. A polar coordinate robot, sometimes called spherical coordinate robot, moves so that the work envelope forms the outline of a sphere. A robot with a cylindrical coordinate system moves so that the work forms the outline of a cylinder.

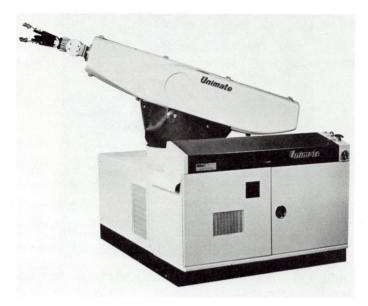

FIGURE 3-15.   The Unimate 2000 robot uses a fixed logic and computer to control the hydraulically driven arm. This robot can move 300 lbs. with a reach of 7.5 feet and a repeatability of ± .05 inch. *Courtesy of Unimation, Inc.*

Simple robots such as those listed as classes 2 and 3 are often used for undemanding tasks. These simple robots are usually very limited in the amount of information that can be stored in their memory. Generally, only sequence and time are used in their programs, although some branching and subroutines are possible with today's low-cost, increasingly powerful programmable controllers. Normally, these devices are restricted to three or four noncontrolled degrees of freedom. Mechanical stops are used on each axis to set the amount of travel. Because they are limited in the number of moves available to the manipulator, simple robots are very dependent on support equipment such as bowl feeders and part presenters. A general rule in robotics is that the higher the intelligence of the controller and the greater the programmability and number of moves of the manipulator and tooling, the less dependent the robot will be on support equipment. Simple robots in classes 2 and 3 will cost anywhere between $5,000 and $20,000.

Medium technology robots have a greater memory capacity and are easier to teach than simple robots, and may be classified in classes 4 and 5. These robots are usually used in single-machine load-and-unload type work, and are not capable of continuous path operations required for welding and spray painting applications. There are many jobs in manufacturing that can be automated by using medium technology robots. Such robots can cost from $15,000 to $35,000 and generally have a repeatability of .050 inch.

Sophisticated industrial robots are considered by many to be the true class of robots and the the leading edge of manufacturing technology. These robots possess highly flexible and programmable manipulators and utilize controllers that exemplify the highest level of artificial intelligence used in industrial automation. Such controllers can be interfaced with sophisticated sensory and inspection devices and also enable the robot to be taught even the most complex of jobs with relative ease. The sophisticated industrial robot has the capability of being integrated into a myriad of computer control work cells and systems. These robots possess a relatively large memory and are capable

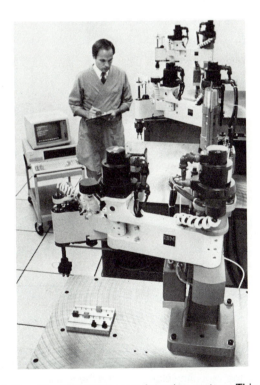

FIGURE 3-16.   The IBM 7535 manufacturing system. This robot has four degrees of freedom with a payload of 6 kg and a repeatability of ± .05 mm (.002 inch). *Courtesy of IBM Corporation.*

of containing multiple programs. They are able to change programs automatically, depending on the requirements of the work cell or product on which they are working. Usually high level programming languages and software are used in conjunction with sophisticated robots. Small sophisticated robots can attain a repeatability of ± .002 inch while carrying loads of a few pounds. Sophisticated robots cost in the range of $40,000 to $150,000, depending on configuration. The IBM 7535 robot is shown in Figure 3-16.

The work space or envelope of a jointed arm robot is shown in Figure 3-17 and the work envelope for the spherical coordinate robot is shown in Figure 3-18.

Robots are increasingly added to a CAD/CAM system and used for material handling, insertion of parts in machines, warehouse control, and automatic quality inspection. Such a robot may be defined as *a reprogrammable multifunctional manipulator designed to move material, parts, tools, or specialized devices through variable programmed*

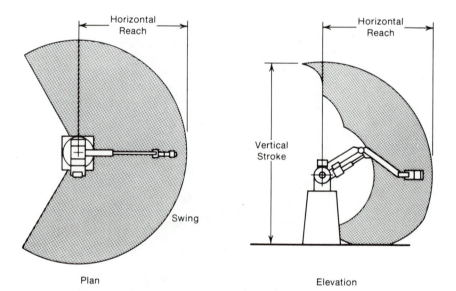

FIGURE 3-17.   The work space (or envelope) of a jointed-arm robot. *Reach* is the maximum distance from the center line of the robot to the end of its tool mounting plate; *swing* is the rotation about the center line of the robot; and *vertical stroke* is the amount of vertical motion of the robot arm from one elevation to the other.

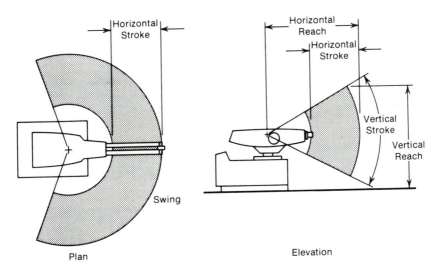

FIGURE 3-18. The work envelope for a spherical coordinate robot.
Horizontal stroke is the amount of horizontal motion of the robot arm.

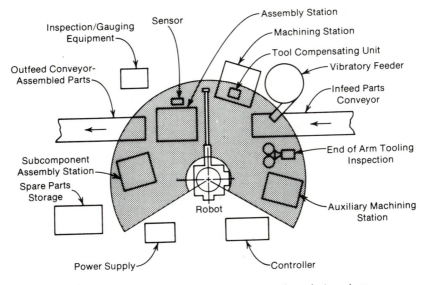

FIGURE 3-19. The potential elements of a robot system.

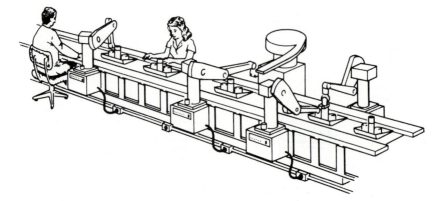

FIGURE 3-20. Robot arms with revolute coordinates may be used alongside humans on an assembly line.

motions for the performance of a variety of tasks. It is the ability of the robot to be reprogrammed that enables the user to rapidly use the robot in new tasks. This is in contrast to the so-called fixed or hard automation which is designed and built for a special purpose and must be discarded when the task is no longer required. The flexibility of a reprogrammable robot enables the robot to have multitask capabilities such as a human worker provides (Vasilash, 1982).

The potential elements of a robot operating in a manufacturing system are shown in Figure 3-19. Another arrangement of workers and robots is shown in Figure 3-20.

The characteristics of a robot may include those listed in Table 3-2. Often several selected characteristics are critical such as repeatability and load carrying capability. *Repeatability* is the ability of the manipulator arm to position the end effector at a particular location within a specified distance from its position during the previous cycle.

TABLE 3-2. Robot Characteristics

Arm Configuration
Number of Axes
Load Carrying Capability
Speed of Movement
Work Envelope
Reliability
Repeatability

TABLE 3-3. Asimov's Law of Robotics

1. A robot may do no injury to a human or allow harm by failure to act.
2. A robot must obey orders from humans except to contradict Law 1.
3. A robot must protect and take care of itself except where contradictory to Law 1 or 2.

Robot Requirements

Issac Asimov has furnished a set of laws which are given in Table 3-3. These laws would restrict robots to actions necessary to the needs of humans and would provide protection of human workers. The robot arm for the NASA space shuttle shown in Figure 3-21, since it is used in cooperation with astronauts, is an example of the necessity of maintaining a well controlled robot.

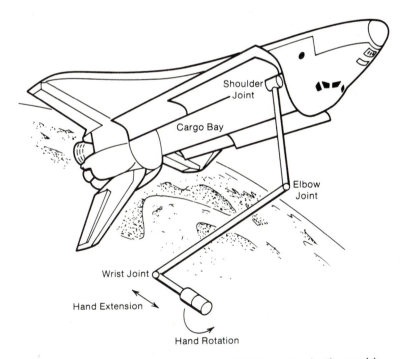

FIGURE 3-21. The robot arm for the NASA space shuttle used in April 1982. The arm has six degrees of freedom including shoulder, elbow, and wrist joints, and hand extension, rotation, and bend.

4

Components and Operation of Robot Systems

The robot system consists of mechanical and electrical components interconnected in order to achieve the desired manipulation and control. The control system for the robot may either be structured as a closed-loop or an open-loop control system; that is, the system may use feedback (closed loop) or it may perform its programmed function without verification of results (open loop). Figure 4-1 shows a feedback system for controlling a single manipulator arm about a single axis. This system utilizes feedback and compares the actual arm angle with the desired input angle Θ_d. The computer calculates the error signal and provides a signal to the power amplifier in order to adjust the motor so that the arm will attain the desired angle.

The single underlying concept of feedback control is that a commanded position, Θ_d, is compared with the actual position, Θ, and the adjusted actual angle is achieved by a motor drive which eventually acts until the error is reduced to zero. Thus, every feedback system consists of three physical components: sensors to measure the state of the system, actuators to do the moving, and a control computer which at minimum must compare Θ and Θ_d and calculate the error. In addition, some form of power amplification and actuation is required (Engelberger, 1980; Dorf, 1981). A comparison of actuator drives is given in Table 4-1.

In general, the control system of a multijoint or multiple axis robot requires several sensors and actuators as shown in Figure 4-2. The thick lines and arrows represent the fact that many signals are present; therefore, the desired position for a robot arm requires a desired angle. The

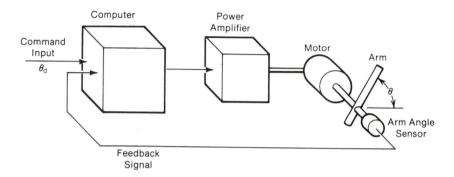

FIGURE 4-1.   A feedback system for controlling a single manipulator arm about one axis.

computer calculates many signals and drives an actuator for each joint. The manipulator then responds, providing the actual output position which is measured by sensors on each of the axes. Again, we must remember that the robot may operate in the open loop configuration; that is, sensors are not utilized and a computer calculates the desired position of the manipulator and gives the signal to the actuators in order to achieve this. Without feedback, there is no verification that the desired position is achieved.

Operating Modes

The Cincinnati Milacron T3 robot shown in Figure 4-3 is a typical six-axis system. Because of its jointed configuration the arm can reach over

TABLE 4-1.   Comparisons of Actuator Drives

	Electric	Hydraulic	Pneumatic
Percent in Use in Robot Systems	25%	45%	30%
Cost	Medium	Highest	Least
Load Carrying Ability	Low	High	Medium
Torque Capacity	Low*	High**	Medium**
Cleanliness of Work Area Due to Fluid Leak, etc.	High	Low	Medium

*Can use gears.
**Can use accumulator.

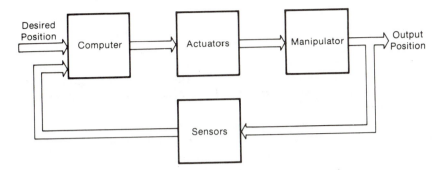

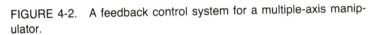

FIGURE 4-2. A feedback control system for a multiple-axis manipulator.

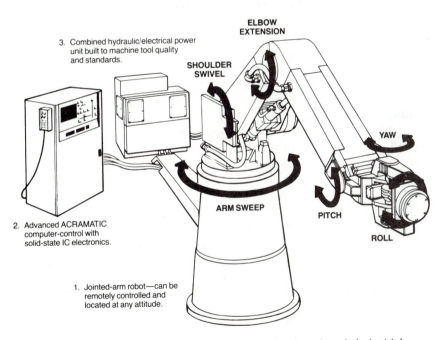

3. Combined hydraulic/electrical power unit built to machine tool quality and standards.

ELBOW EXTENSION

SHOULDER SWIVEL

YAW

ARM SWEEP

PITCH

ROLL

2. Advanced ACRAMATIC computer-control with solid-state IC electronics.

1. Jointed-arm robot—can be remotely controlled and located at any attitude.

FIGURE 4-3. The Cincinnati Milacron T3 robot is a six-axis industrial robot using electrohydraulic servo control systems to achieve high torque and speed. This system provides a controlled path for a 100 lb. load with a repeatability of ± 0.05 inch (± 1.27 mm). *Courtesy of Cincinnati Milacron.*

150 inches (380 cm) vertically upward and extend horizontally almost 100 inches. The control computer is a 32K byte minicomputer. Up to 700 programmed points can be stored in memory. Controlled path operation refers to the ability of the system to provide coordinated control of all axes during both the *teach mode*—the process of programming a robot to perform a desired sequence of tasks—and automatic operation. During teaching, the operator guides the robot to the desired locations by commanding a direction of motion for the tool (such as up, left, or out), as opposed to directing an individual axis. The operator is not required to generate the required path during teaching, but only to identify individual path end points. During replay and automatic operation, the computer control directs the arm along a straight line path between the programmed end points, at a specified velocity, with acceleration and deceleration signals provided.

The Tool Center Point (TCP) is a point that lies along the roll axis a user-specified distance from the roll face plate. It is the coordinates of unique locations of the TCP that are programmed as the robot is taught its assigned task. Also, it is this point that follows the straight line path as the arm moves from one programmed location to another. Furthermore, all orientation of the wrist axes takes place about the TCP.

As discussed previously, during teaching the operator depresses the pendant position and orientation buttons to cause the TCP to move in a coordinated manner in the teach coordinate system selected (rectangular, cylindrical, or hand). As long as the buttons are depressed, continuously changing position signals are generated. The *teach coordinates* are then transformed into *rectangular coordinates* in the first operation performed within the computer. These coordinates are, in turn, transformed into *robot axis coordinates*—values that tell each axis actuator how much to move. Such values are then output to the axis servo system as command signals. The six servo loops drive all the axes simultaneously to provide the desired change in position and orientation of the end effector.

When the PROGRAM button on the pendant is pushed, the rectangular coordinates of the current TCP and current wrist orientation angles are placed in the computer's memory. To these data are added the selected path velocity for getting to this point, the function to be performed at the point, and the tool length distance from the face plate to the TCP. These data constitute what the arm needs in the automatic mode.

During the *automatic mode*, the information stored in memory is recalled as needed. This information is sent through a path generation algorithm which computes incremental locations along a straight line path between two programmed points. These intermediate points on the path are generated in a proper time sequence to cause the TCP to move at the programmed speed and also generate the proper acceleration and deceleration spans. These incremental locations are then transformed into robot (joint) coordinates before being output to the axis servos. Thus, the path generation is an on-line calculation done in realtime, so that the computer always knows the current location of each axis of the arm.

Operation of the T3 computer control system is shown in Figure 4-4.

The latest robot from Cincinnati Milacron is the T3R3, which has the same base, shoulder, and upper arm as Milacron's T3 robot but incorporates a new wrist design. The wrist on the T3R3 is shaped like a sphere, with a diameter of about 20 cm, and it consists of three roll axes coincident at a single point. The two roll axes that are closest to the robot forearm, working in combination, provide 230 degrees of pitch (up and down) motion and 230 degrees of yaw (side to side) motion. The third axis (farthest from the forearm) provides continuous roll for rotation of the tool. The wrist contains three concentric shafts that drive the axes and are powered by the hydraulic motors mounted at the elbow of the robot's arm. The light weight of the wrist provides low inertia for each wrist axis, and the remote drive system provides enough stiffness for satisfactory servo control. Also, each axis has its own feedback device, consisting of a resolver and tachometer for precise positioning. The standard positioning accuracy of the robot is ±0.5 mm. For a robot, *accuracy* is defined as *the ability of the manipulator to position the end effector at a specified point in space upon receiving a command by the controller.*

The control and components of a robot within the total manufacturing system are chosen so as to achieve a high degree of reliability. It is possible to arrange the work place in four different configurations:

1. The work can be arranged around the robot.
2. The work can be brought to the robot.
3. The work travels past the robot.
4. The robot travels to the work.

In any case, there is some form of arrangement between work place and robot in order to achieve a hierarchy of control tasks in an optimum configuration. The one possible arrangement for hierarchical control of a robot is shown in Figure 4-5. In this form the system controls many work stations and the work station controls the robot as well as the machine, tools, or other devices. Within the robot the control is on the elemental movements and then passes to trajectory control and finally

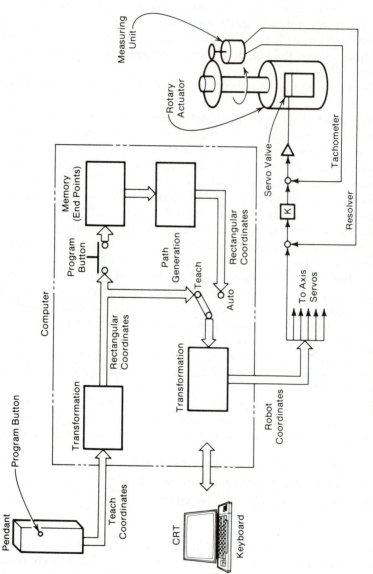

FIGURE 4-4. T3 Computer Control System.

68

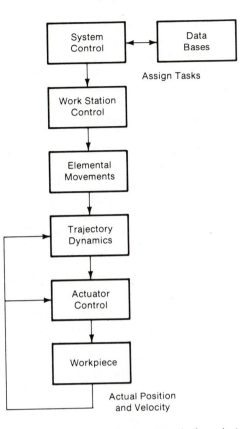

FIGURE 4-5. Hierarchical control of a robot.

to actuator control. The work place achieves the actual position and this information may be fed back to the actuator and the trajectory calculation, as shown in Figure 4-5 (Saveriano, 1980; Dorf, 1981).

Control System Sophistication

One way to look at the robot is to describe it as consisting of two major systems. The first system comprises the mechanical components including the arm, the wrist, and the end effector. This is the part of the robot obvious to the observer. In addition, the robot depends on the control system as described earlier. At its simplest, this control system can consist of a series of adjustable mechanical stops or limit switches. This simple robot often is called a *pick and place* device or a *limited sequence* robot. Simple robots are often underrated and under utilized.

These low-cost, easy-to-maintain, fast and accurate devices can dramatically improve productivity in medium production industries. An example of a simple limited sequence robot is shown in Figure 4-6.

Usually, only the sequence and time of dwell at each point are stored in the programs, although some branching is possible utilizing programmable controllers added to the simple robot. Normally, these devices are restricted to three or four axes and do not utilize feedback but rather operate on an open loop control. Mechanical limit switches or stops are used on each axis to set the amount of travel, and there are usually only two positions on each axis, for example, up or down. Simple robots usually are pneumatically operated and accurate to ±.001 inch and can operate as fast as one cycle per second.

The limited sequence system is capable of taking the arm from point *a* to point *b*, but the path in between is not defined. Thus, the control simply switches the drives on and off at the ends of travel— hence the term *limited sequence* robot. The drive mechanism could be electrical, pneumatic, or hydraulic, but most robots of this type utilize mechanical stops and pneumatic drives. The disadvantage of these systems, other than the obvious control limitations, are that the number of limb articulations is likely to be few and setting the machine up is more time consuming and tedious than for those with more extensive control systems. The number of movements possible in a total production sequence must be limited to the number of limit switches, stops,

FIGURE 4-6.   The Seiko robot moves linearly in the *x* and *z* directions (in/out and up/down) using pneumatic drives. *Courtesy of Seiko Corporation.*

and programmable switches contained by the robot. Such robots are set up in the same way that one would throw a series of switches to desired positions. The memory consists of the logic established in relays or electronic logic and all the mechanical stops. The controller of such a system is a sequencing or stepping switch. It can be a set of contacts operated by CAM's on a spindle which can be rotated in steps of a few degrees at a time by a small electric motor. Alternatively, the sequence can be achieved by an electronically driven clock and a series of logic gates.

The alternative to point-to-point control is *continuous* or *trajectory control*, whereby the trajectory or continuous control is achieved by some form of automatic control system, as shown in Figures 4-1 and 4-2. Continuous feedback then enables the operator to guide the robot over a continuous path between point *a* and point *b*. If the desired continuous path is known, one form of operation is achieved by using a memory which stores the desired position over the trajectory and thus is used to drive the actuators. An example of a sophisticated robot capable of continuous path control is the Cincinnati Milacron T3 robot shown in Figure 4-3. As described earlier and shown in Figure 4-4, it is possible to teach the robot the desired path of control. This is achieved by storing the desired path in memory while leading the robot through the desired path. Then the robot is switched to automatic operation and the memory is used to drive the robot arm. Sophisticated systems capable of continuous trajectory control utilize relatively large memory systems capable of containing multiple programs and have the ability to change programs automatically depending on the requirements of the work place.

It is possible to take the continuous path robot and teach it in real time. In order to achieve this, the operator takes hold of the robot at its end effector and, attempting to copy the speed of travel desired, leads it through the motions it is to perform. During this teaching process, the robot has to record the movement and hand attitudes continuously. This is achieved by giving the robot an internal timing system through which the movements are synchronized.

Actuator Drive Systems

An actuator drive system is required for each robot articulation. In addition to driving the arm, the hand, and the wrist, most types of grippers need a drive mechanism for the functions of holding and release. Robot drives can be electrical, pneumatic, hydraulic, or some combination of these.

Pneumatic systems are found in about 30% of robots, although they are confined mainly to the limited sequence devices. Pneumatic drives have the merit of being cheaper than other methods, and their inherent reliability means that maintenance costs can be kept down (although other types have also become reliable). Since machine shops typically have compressed air lines available throughout their working areas, this makes the pneumatically driven robot a more familiar tool to shop personnel. Unfortunately, the system does not make for easy control of either speed or position, essential ingredients for any successful robot.

Electromechanical drives are used in some 20% of robots. These systems are servo motors, stepping motors, pulse motors, linear solenoids, and rotational solenoids. But the most popular form of drive system for large loads is hydraulic, because hydraulic cylinders and motors are compact and allow high levels of force and power, together with accurate control. An *hydraulic actuator* converts forces from high pressure hydraulic fluid into mechanical shaft rotation or linear motion. While electromechanical drives have taken over some functions, such as precision feed drives, fluid power is still more cost-effective for short-stroke, straight-line positioning requiring high forces, controlled acceleration, and repetitive motion. No other drive packs as much power in as small a package; no other drive is as inherently safe or as resistant to harsh environments.

Short-stroke, straight-line positioning is required for automatic tool changers on modern machine tools. The hydraulic system can supply tremendous forces to hold both the tool and the work piece in position under high cutting forces (Beercheck, 1982). A fluid-power drive can deliver high power directly where it is needed, rather than through an intermediate member. This, combined with fluid power's inherent compactness, minimizes the weight on the robot arm, thus increasing payload. A fluid-power drive for a robot is also simple and durable.

Pneumatic robots typically are limited to light-duty applications involving the transfer of parts weighing less than 50 pounds. Hydraulic robots handle much heavier loads and are used to load and unload machines and to perform repetitive operations such as spot welding and painting. A low-power pneumatic robot is shown in Figure 4-6.

A key element in nearly every robot is the servo valve system. This system translates the computer position commands into motions of the fluid-power actuators and cylinders. On Cincinnati Milacron's HT3 robots, for example, each of the six arm axes is controlled by its own closed-loop servo system. Five axes are driven by hydraulic rotary actuators; the sixth by a pivoted cylinder. The servo system provides precise, repeatable arm positioning, and the hydraulic system provides the high torque and speed needed to pivot 225 lb. loads through 270 degrees.

A comparison of actuator drive characteristics is given in Table 4-1. The electric drive systems are particularly useful for highly accurate, lightly loaded systems. Figure 4-7 shows the range of response time and power for electromechanical and electrohydraulic actuators. As shown in the figure, an electric drive is used for lower power applications. An example of a low-cost robot using electric motor drives is shown in Figure 4-8.

Another common electric motor drive uses stepping motors. Because it is possible to design very simple control systems for stepping motors, they are being used in a wide variety of applications. Their diverse applications include machine tools, process control, computer peripheral equipment, and, recently, robots. In general terms, a *stepping motor is an electromagnetic incremental actuator which converts digital pulse inputs to discrete motion steps.* In the familiar rotary stepper, the shaft rotates in equal increments in response to an appropriate input

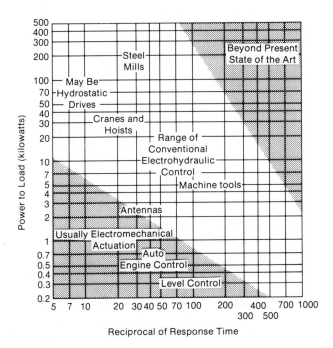

FIGURE 4-7.   Range of control response time and power and load for electromechanical and electrohydraulic actuators. *Source: R.C. Dorf, Modern Control Systems, Addison-Wesley, 1981.*

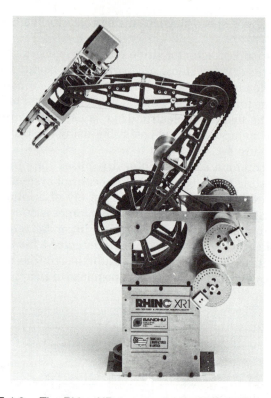

FIGURE 4-8.   The Rhino XR-1 computer-controlled robot arm incorporates six DC servo motors and uses gears, cables, and chains for the drive linkages. Optical digital encoders are used on each axis and the system achieves a horizontal reach of 22.25 inches. *Courtesy of Sandhu Machine Design Inc.*

pulse train. There are a variety of stepping motor designs, the most popular of which is the permanent magnet rotary type.

A stepping motor has several operating modes, depending upon the stepping rate and load conditions under which it operates. If the stepping motor is pulsed at a sufficiently slow rate, it comes to rest at the end of each step, moving in discrete steps. In applications where the distance to be traveled is small and the response time is not critical, this mode of operation provides the simplest solution (Motiwalla, 1981).

As the input pulse rate to the motor is increased, the motion changes from discrete steps to a continous motion referred to as *slewing*. In the transition between these operating modes, the motor does not come to rest between steps; thus subsequent switching pulses may come when the motor has either a positive or negative velocity. At low torque in the discrete stepping mode, the load on the motor is low enough for it to step at the given rate without missing any pulses and with the motor returning to rest after each step. Operating in the slewing mode, however, the motor can produce greater torque at a given step rate, but changes to the step rate must be made carefully so that the peak torque is not exceeded. Above the peak torque, the motor may fail to step in response to an input pulse. One of the key advantages of using a stepping motor is that, by observing its torque/step rate limitations, you can use the motor "open loop," with full confidence that you can tell where its shaft is (relative to some initial position) merely by keeping track of the number of steps sent to the motor.

A small, low-cost robot, the Mini Mover 5 shown in Figure 4-9, uses stepping motors and a cable drive system to achieve control of a jointed arm (Hill, 1980). Cable drive was selected because it is lightweight, highly efficient, and permits all the drive motors to be mounted on the body. This keeps weight out of the extremities, resulting in a payload capacity of 225 gm (8 oz.). The shaft of the base joint is hollow, so that the drive lines for each of the motors can be brought from the interface card contained in the base of the arm. Since the axes of the three outer joints are parallel, mounting the drive motors on the body instead of the base eliminates the complications of routing the drive cables for the outer segments through a rotating joint. The drive unit

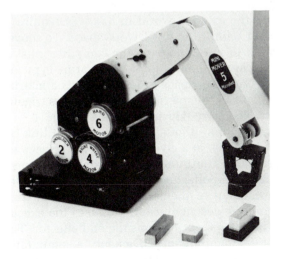

FIGURE 4-9.   The Mini-Mover 5 robot uses six stepping motors and a cable drive system. *Courtesy of Microbot Inc.*

for each joint consists of a stepping motor, reduction gearing, and a cable drum. From each drum, a tensioned cable goes out over pulleys to the member being driven and then returns to the drum. Rotation of the drum causes rotation of each member in proportion to the ratio between the diameter of the drive pulley attached to that member and the diameter of the drum.

Kinematics of Robot Arms

While the dynamic equations for a robotic manipulator can be developed using various methods that are different in form, the predicted dynamic behavior of the manipulator will be the same. In each method the resulting behavior is given by a highly nonlinear set of coupled ordinary differential equations. For example, manipulators may be analyzed using Lagrange's equation.

The kinematic configuration of a system is typical of a large class of industrial manipulators. For the purposes of analysis, the manipulator's elements are assumed to be rigid; effects such as connection clearances are neglected. The joint angles and angular velocities are assumed measurable, for control purposes.

There are two principal ways to specify a manipulator path: as a set of hand positions and orientations, or as a set of joint angles. Which representation is most appropriate depends on the task. In the most general of robot tasks, the robot must track and acquire some object moving in the robot's working area, or apply a controlled force. To coordinate with the external world in this way requires the ability to control the hand within the fixed work space. To move in this way, given the desired location in space, the computer must determine the proper torque to apply to each joint of the arm. Typically, this transformation from Cartesian space to joint space is a matrix operation in which the elements of the matrices are themselves trigonometric functions of the arm geometry (Huston & Kelly, 1982).

The determination of joint angles for a multilink manipulator to position its end effector at a given point in work space is known to be a complex problem. The number of possible solutions for the joint angles to reach an end effector position increases with the number of links. The consequent mathematical complexity of the algorithm to compute the joint angles significantly reduces the execution speed of the computer program that positions the end effector. To achieve real-time control of the manipulator movements, a computationally simple algorithm to determine the joint angles for a given end effector position is required.

There is considerable discussion in the technical literature concerning the optimal procedure for obtaining the equations of motion for robotic and other complex mechanical systems. Some investigators have suggested that Lagrange's equations present the most convenient and efficient procedure for developing the governing equations. Others have maintained that there are distinct computational advantages with using Newton's laws. Still others have asserted that neither a strict Lagrangian or Newtonian approach is optimal, but that a modified combination is preferred. Results of these discussions are likely to be inconclusive unless the characteristics of an optimal procedure are commonly accepted. However, most investigators are currently seeking methods which: (a) are easy to formulate, (b) are easily converted into computer algorithms and codes, and (c) lead to efficient numerical solutions (Huston & Kelly, 1982; Paul, 1981).

The challenge of calculating the correct equations of dynamics can be illustrated by the Canadarm shown in Figure 4-10. This arm was used on the U.S. Space Shuttle during 1982.

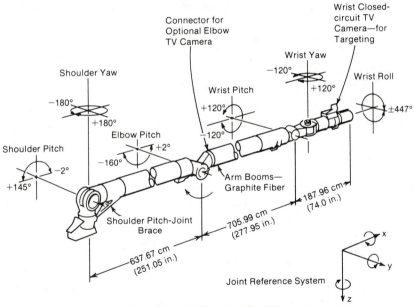

Note: Joint Angles Show Specified Values

FIGURE 4-10.   The robot arm used on the NASA space shuttle. The arm, called Canadarm, weighs 905 lbs. and can move large payloads in gravity-free space. The arm consists of graphite-epoxy booms, uses electric motors, and can reach 600 inches when fully extended.

End Effectors

The end effector is at the action end of the robot; it is the tool or gripper which is attached to the mounting surface of the manipulator wrist which has to grasp, lift, and manipulate work pieces without causing any damage and without letting go. Compared with human hands, those of the robot are very limited. They have fewer articulations and they are without any sense of feeling or touch. However, they can be designed to withstand high temperatures so that they are able to work with parts that are red hot, and they are also better at dealing with objects which have sharp edges, are covered with corrosive substances, or are simply too heavy for human hands to lift.

End effectors have to be chosen or designed specially for a particular industrial application. Whereas the robots themselves have earned the reputation of being general-purpose automation, the effectors are not quite so flexible and may have to be included along with the special tooling requirements of the job. End effectors are often called grippers, although this implies only one form of effector. One type of effector is usually suitable for a wide range of different jobs at a particular work station. Only when the robot has to be redeployed elsewhere to work on an entirely different process is it likely that the hand tooling has to be changed. And, compared with overall plant and machinery costs, effectors come relatively cheap.

There are several ways of grasping or gripping a tool or piece. These include:

- mechanical grippers
- hooks
- a thin platform or spatula
- scoops or ladles
- electromagnets
- vacuum cups
- fingers using adhesives
- bayonet sockets

In addition, an end effector is particularly useful if it can change from one tool to another using an adjustable connector.

A sample cam-driven gripper is shown in Figure 4-11a. If an application calls for a gripper to remove a finished part and replace it with an unfinished part, a two-gripper end effector can be used as shown in Figure 4-11b.

The MiniMover gripper and wrist is shown in Figure 4-12. This

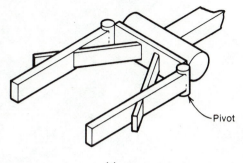

(a)

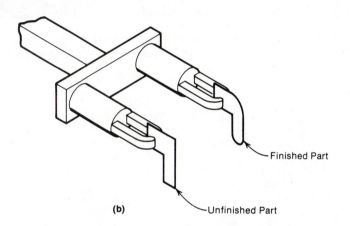

(b)

FIGURE 4-11.   (a) A simple cam-operated gripper. The fingers rotate at the pivot. (b) A two-gripper end effector.

system uses a differential gear to rotate the wrist. A cable can pull the gripper together. An example of a vacuum gripper is shown in Figure 4-13, and a specially designed gripper is shown in Figure 4-14.

Compliance

The term *compliance* refers to *the yielding or deformation of a component due to externally applied forces and inertia.* For example, a

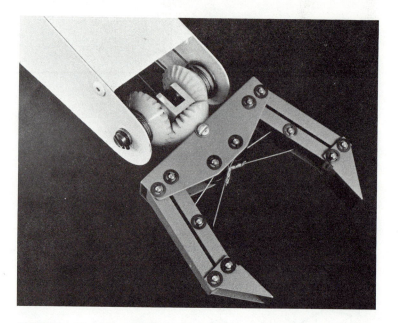

FIGURE 4-12.   The MiniMover 5 wrist and gripper. A differential gear in the wrist enables rotation. The gripper cable closes the fingers. *Courtesy of Microbot Inc.*

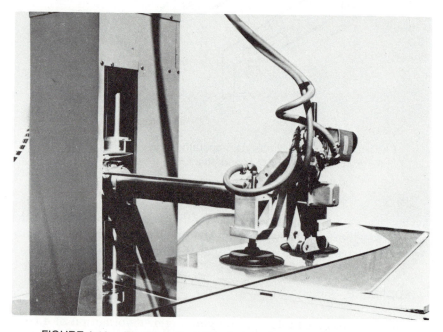

FIGURE 4-13.   The Prab Model E robot stacks automobile windshield glass using vacuum grippers. *Courtesy of Prab Corporation.*

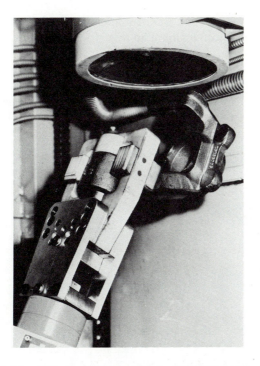

FIGURE 4-14. This specially designed gripper provides extremely accurate positioning for testing purposes. *Courtesy of Reis Machines Inc.*

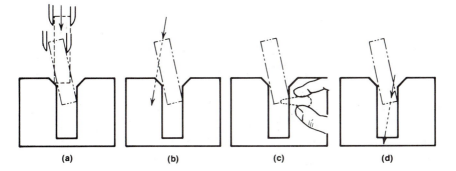

(a)   (b)   (c)   (d)

FIGURE 4-15. Compliant grippers used to insert a peg into a hole.

robot wrist will experience a compliance force when exerting effort to insert a work piece into a slot. The less stiff (more compliant) the parts and the grippers are, and the lighter the arm's moving components are, the easier it is to obtain rapid, stable, and effective responses with low-contact forces at the tip of the part. When low stiffness and rapid response motion cannot be built into the apparatus (because, for example, it is too heavy or the work piece it is holding is), the only remedy

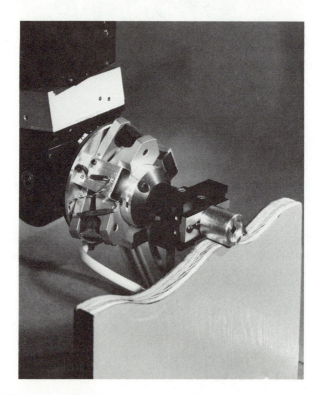

FIGURE 4-16.   This photo shows an Instrumented Remote Center Compliance (IRCC). The IRCC and a gripper are attached to a robot arm, partly visible at the left. The instrumentation allows measurement of the position and angle displacements of the tool or gripped part relative to the robot's wrist. The compliance is achieved by means of elastic shear pads shown on the wrist. The IRCC can accommodate misalignment errors. Combined with control algorithms, IRCC robot systems can automatically learn the shape of a contour (shown in photo), track a seam for welding purposes, put a peg in a hole when neither has any chamfers, or gather statistical tactile data on manufacturing operations such as assembly, welding, grinding, and deburring. *Courtesy of Charles Stark Draper Laboratory Inc.*

for avoiding large contact forces is to make all the close-loop motions slowly. This alternative is an unattractive one from an economic point of view. A conclusion of recent studies shows that devices capable of fine assembly motions must be small, light, and fast (Nevins & Whitney, 1978).

In many ways assembly is best understood in terms of the forces and moments acting on the tip of the part, where it touches its mate during assembly. For example, if compliant grippers are used to hold a peg, a lateral error becomes an angular error as the peg slides down the chamber and enters the hole as shown in Figure 4-15a. Continued application of force at the top of the peg (Figure 4-15b) creates a torque that can lead to jamming. If the peg could be grasped compliantly at its tip (Figure 4-15c), insertion could be accomplished in spite of substantial angular error. The same force that would cause jamming if it were applied at the top of the peg would tend not to cause jamming if it could be applied at the bottom of the peg (Figure 4-15d). Grippers

FIGURE 4-17. The Cybotech P15 is a hydraulically powered robot capable of 2 m/sec (6.6 ft/sec) with a 15 kg load. Repeatability is ± 5 mm (± .20 inch). This six-axis robot can be used for painting and coating applications, among others. This robot is manufactured by Cybotech Corporation, a joint venture company owned by Ransburg Corp. of Indiana and Renault Industries, France. *Courtesy of Cybotech Corp.*

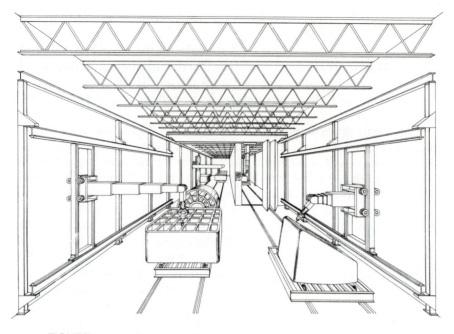

FIGURE 4-18.   A concept of a wall-mounted, extended-reach series of robots. The extended reach is achieved by utilizing two tracks on which the robot rides horizontally and vertically. *Courtesy of Par Systems, GCA Corporation.*

with compliance can correct for both lateral and angular misalignment in the same way a human hand and wrist are able to respond. A compliant physical member such as a robot wrist must be flexible enough to allow correction for misalignment. An example of a device utilizing compliance is shown in Figure 4-16. This device can learn the shape of a contour or track a seam for welding.

Robot Design

Robots can be designed to fit into large industrial systems. They will consist of some form of manipulator and end effector. An example of a hydraulically driven painting robot is shown in Figure 4-17. In addition, robots can be combined in a form of manufacturing system using both machines and robots. An example of a design for a wall-mounted, extended-reach system combined with tracked vehicles is shown in Figure 4-18. One of the earliest robot manipulators designed for artificial intelligence research is shown in Figure 4-19.

FIGURE 4-19.   The Stanford University manipulator arm, an example of a manipulator using a polar coordinate configuration. This six-axis arm was designed and built by Victor Scheinman in 1969. The arm rotates at the base and shoulder and moves linearly at the forearm. There are three axes of movement at the wrist. *Courtesy of Stanford University.*

5

Types of Robotic Sensors

To perform some of the tasks presently done by man, a robot must be able to sense both its internal state and its environment. A *sensor* is defined as *a measurement device which can detect characteristics through some form of interaction with them.* A robot sensor consists of a transducer, which converts physical properties into a signal (e.g., electrical), and a processor, which transforms the signal into the information required to perform a given task (Rosen & Nitzan, 1977).

Only rudimentary sensors (e.g., a microswitch) are currently applied to robots on factory floors, and this fact reduces their flexibility, accuracy, and repeatability. However, newly developed sensors, especially visual sensors, and computer control will soon be incorporated into industrial robots.

In order to be able to use sensors, one must recognize the various types of sensors available and their main properties. This chapter describes the main types of industrial sensors and brings some examples of their utilization with robots. In addition, some experimental sensors are mentioned. Vision, which is an important sensing capability, is a broad subject by itself and will be covered in the next chapter.

Classification of Sensors

Sensors for robotics can be classified in different ways, such as contact or non-contact, internal sensing versus external sensing, passive versus active sensing, and others. In this chapter, the sensors are basically classified in two groups: contact and noncontact sensors. In each group a further classification is made according to the physical properties that are sensed.

A noncontact sensor measures the response of a target to some form of electromagnetic radiation (visible light, x-ray, infrared, radar, acoustic, electric, and magnetic radiation). A contact sensor, on the other hand, measures the response to some form of physical contact (e.g., touch, force/torque, pressure, position, temperature, electrical and magnetic quantities). The contact sensors group consists of devices which can sense touch, force, pressure, tactility, and temperature. Some of these sensors can replace vision in cases where they are accurate enough and vision is not suitable because of an environment without illumination or the property cannot be measured by vision (Rosen & Nitzan, 1977).

Range Sensors

A range sensor measures the distance from a reference point (usually on the sensor itself) to a set of points in the scene. Humans can estimate range values based on visual data by perceptual processes that include stereopsis as well as comparison of image sizes and projective views of world-object models. Some animals (e.g., the bat and the dolphin) can estimate range values by use of active range sensing in which a sonic wave is transmitted and the elapsed time for the return echo is calculated.

Basic optical range-sensing schemes are classified according to the method of illumination (passive or active) and the method of range computation (triangulation or time of flight of light). Range can be sensed with a pair of TV cameras or sonar transmitters and receivers. Range sensing based on triangulation has the drawback of missing data for points not seen from both positions of the transmitters. This drawback may be reduced, but not eliminated, by using additional transmitters. The use of additional cameras may also provide a partial solution to the general problem of occlusion, including self-occlusion, in machine vision. Instead of a projector and camera scheme, one system has a single projector with multiple TV cameras or camera front-ends, viewing the target from different angles. Such a solution also entails additional computer processing. Justification of the additional hardware and software will depend on the importance of the additional information this scheme provides.

The main drawback of the laser-scanner and photomultiplier ranging scheme is that it is too slow, especially if the target is dark. This drawback may be resolved by increasing the laser power, increasing the photomultiplier-receiver area, and improving other sensor parameters.

Acoustic rangefinders, like the one used in the Polaroid cameras, yield only a single range value. To obtain a range image, the object must be scanned and spurious echo signals must be disregarded. In addition, for high spatial resolution new techniques are needed to overcome the absorption of the energy of a high-frequency acoustic wave by its medium.

Range sensing has only recently been utilized in performing robotic tasks, such as object recognition and inspection, manipulation, and navigation. More research is needed in this area for such applications. Simultaneously, research and development effort is also needed to improve the capabilities and to reduce the cost of range sensors.

Proximity Sensors

A proximity sensor senses and indicates the presence of an object within a fixed space near the sensor without physical contact. Different commercially available proximity sensors are suitable for different applications. For example, eddy-current sensors can be used to precisely maintain a constant distance from a steel plate. A common robotic proximity sensor consists of a light-emitting-diode (LED) transmitter and a photodiode receiver. The main drawback of this sensor stems from the dependency of the received signal on the reflectance and orientation of the intruding object. This drawback can be overcome by replacing proximity sensors by range sensors.

The proximity sensor usually detects a disturbance in the nearby field which is caused by the presence of the object. The difference between various kinds of proximity switches is due to their sensitivity to different types of fields, and the different methods that are used to sense a disturbance in the field.

A reed relay is a simple proximity sensor which is activated by a magnetic field. The reed relay is contained in a vacuum glass tube and is made of a ferrous material. When an external magnet is placed near the reed relay, the magnetic lines of the field concentrate in the arms of the relay; in reacting to shorten the magnetic path, they approach each other and close the contacts.

Other proximity sensors include Hall effect sensors and eddy current sensors.

Acoustic Sensors

An acoustic sensor senses and interprets acoustic waves in gas (non-contact sensing), liquid, or solid (contact sensing). The level of sophistication of sensor interpretation varies a great deal among existing acoustic sensors, from a primitive detection of the presence of acoustic waves, to frequency analysis of acoustic waves, to recognition of isolated words in a continuous speech.

Animals utilize natural acoustic sensing for detection of events, communication, and other functions, and although man has utilized artificial acoustic sensing to augment similar functions, it has only recently been applied to robotics. This situation will change as the application of robots increases. For example, in addition to man-robot voice communication, acoustic sensing can be utilized by robots to assist in controlling arc welding, to stop the motion of a robot when a loud crash is sensed, to predict a mechanical breakage about to happen, to implicitly or explicitly inspect objects for internal defects, and so forth. Research is needed to develop methods and software for successful use of acoustic sensing in such applications.

Temperature Sensors

Temperature sensing, both contact and noncontact, also has only recently been performed by robots. Such performance is useful where robots operate autonomously, where human presence is undesirable or where temperature images are required. There is a need to increase the accuracy of hot temperature sensors (e.g., for measuring the temperature of molten steel) and to improve their area imaging capability. Significant progress in sensing temperature images has been achieved in recent years by improving the capabiliites of pyroelectric television cameras.

Two common types of temperature sensors are the thermistor and the thermocouple. Both must be in physical contact with the object whose temperature is being measured. The thermistor is a special resistor which changes its resistance proportional to the temperature. The thermocouple is a device which produces a small voltage that is proportional to the difference between two temperatures. In order to use a thermocouple, usually one part of it is connected to some temperature reference and then temperatures relative to the reference can be measured.

Touch Sensors

A touch sensor senses and indicates a physical contact between the object carrying the sensor and another object. A simple touch sensor is a microswitch. Basically, touch sensors are used to stop the motion of a robot when its end effector makes contact with an object. Such control is applicable to a variety of tasks, including:

- Reaching a target
- Preventing collision
- Centering the robot grippers on an object without moving
- Object recognition

Force Sensors

A force sensor measures the three components of force and three components of torque acting between two objects. In particular, a robot-wrist force sensor measures the components of force and torque between the last link of the robot and its end effector by transducing the deflection of the sensor's compliant sections, which results from the applied force and torque.

Existing force sensors employ different transducers such as a piezoelectric element or strain gauges. The best transducers for robots are semiconductor strain gauges cemented onto the compliant sections. Future improvements should result in reduction of size, weight, and cost, and increase in accuracy, resolution, and dynamic range of force sensors.

As with touch sensing, there is a need to extend force sensing from a single point to an array of points of high spatial resolution (e.g., 1 mm). Such an array force sensor could be used to determine the identity, state, centroid, and orientation of an object resting on the sensor where visual sensors are inappropriate. Mounted on the grippers of a robot, two such array force sensors can be used to verify that the proper object is held at the proper gripping locations with the proper force and that no slippage has occurred. As with touch sensors, a local microprocessor should analyze the sensed data to eliminate interface wiring and relieve the higher level computer from force/torque data processing.

Tactile Sensors

Human workers effectively use their ability to sense the presence and outline of an object with the sense of touch. Researchers are also developing artificial tactile sensors for robots. Whereas vision may guide the robot arm through many manufacturing operations, it is the sense of touch that will allow the robot to perform delicate gripping and assembly. Tactile sensors will provide position data for contacting parts more accurately than that provided by vision. And in some cases, the robot may be able to feel its way through a task in regions obstructed from view (Harmor, 1982).

Most robot grippers presently have switches to indicate the presence or absence of an object. Also, grip-controlling sensors detect how tightly an object is grasped. However, sensitive tactile systems are being developed for handling parts more delicately. The simplest type of tactile sensor is a gripper which is equipped with an array of miniature microswitches or a calculator-type array of switches as shown in Figure 5-1.

In one advanced approach, skin-like arrays of pressure-sensitive piezoelectric material on the robot gripper generate small electric currents when stressed. This current is applied to a built-in microprocessor and analyzed to yield tactile information about the object's shape and the grip pressure applied. Thousands of these tiny sensors detect subtle

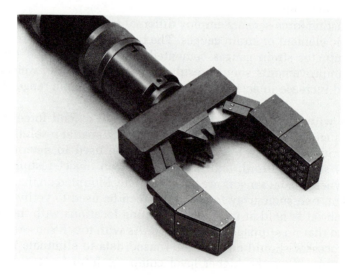

FIGURE 5-1. A tactile sensor using an array of switches mounted on a gripper. *Courtesy of SRI International.*

pressure changes and a multifingered gripper accurately responds to computer commands based on the tactile feedback. This type of system operates in a manner similar to that of the human tactile system, in which electrical nerve impulses are routed via a nerve network between the fingers and the brain.

There are three major potentially desirable properties of tactile sensors:

1. Tactile sensors should be skin-like; that is, they should be distributed in arrays on thin, flexible substrates. The substrate should be compliant, and the entire structure very durable.

2. The sensing devices should be hand-like, having flexible, jointed fingers of great dexterity.

3. The hands must be intrinsically intelligent, with a fair amount of preprocessing done at the sensor level. This can be accomplished by distributed-logic arrays and processed (or multiplexed) so that only a very few output signal paths are needed to go to the central processing unit.

A frequently-cited need is for intelligent hands in tooling operations. This includes the widespread secondary operations in manufacturing which include machine loading and unloading, and removal of pieces from bulk and insertion into other machines for flash removal, tapping, and deburring. These follow on to the primary operations of casting, forging, and stamping, for example, and are conventionally labor-intensive. It is in this domain that sensors on objects such as machines and tables as well as on end effectors may prove to have great utility.

Use of area touch sensors to recognize objects on the basis of touch patterns is presently difficult because these sensors have insufficient compliance and coarse spatial resolution. Coarse resolution, in turn, is caused by the bulkiness and high cost of existing touch transducers. To overcome these limitations, there is need for research and development on a high-resolution compliant array of touch transducers.

The more conventional uses of tactile sensors are: factory assembly, fast adaptive grasping (e.g., pickup of arbitrarily oriented parts from assembly lines), bin picking, grinding, deburring, polishing, welding, undersea applications, and space assembly. Other uses are: recognition in the dark, flexible materials handling, replacement of conventional wrist/arm sensors, circuit-board assembly, force sensing for lead bonding in integrated circuits, process verification (including measurement), rivet insertion, bolt-thread pickup sensing, determination of weight and center of inertia, geological prospecting, paraplegic limb feedback, and fire fighting. It may be observed that, in general, robotics may well be considered more frequently for sensing and surveillance tasks which require few mechanical effecting tasks.

Many observors foresee tactile sensors replacing vision sensors. Another view is that the two are naturally complementary and should be so designed and used. Vision is conventionally considered to be a requisite early process to locate, identify, and position the robot, and to act as a proximity sensor. Tactile sensors are thought of as then taking over for subsequent manipulations where force, pressure, and compliance are important variables. Visual sensing, of course, can continue usefully while touch proceeds, but it is considered to be far less important during manipulation for many applications.

The specifications of a touch-sensing transducer could be:

1. A typical array of 10 × 10 force-sensing elements on a one-square-inch flexible surface, much like a human fingertip. Finer resolution may be desirable but is not essential for many tasks.

2. Each element should have a response time of one to ten microseconds.

3. Threshold sensitivity for the element ought to be one gram, with the upper limit of the force range being 1,000 grams.

4. The elements need not be linear, but they should have low hysteresis.

5. This skin-like sensing material has to be rugged, standing up well to harsh industrial environments.

The state of the art of touch and tactile sensing technology of robotics is in its early infancy. Industrial and other commercially available manipulators having touch capability are crude devices using simple limit switches, potentiometers, and photoelectric sensors. Some other primitive, but useful, force-sensing systems monitor air pressure or electric currents signals. Most provide binary signals only.

However, new approaches are under development. Multijointed fingers and hands are beginning to employ pressure-sensitive conductive material. Simple point-sensing arrays (3 × 3, 4 × 4, etc.) showed early success in discriminating generic geometric shapes by touch. Other lab developments indicate that the same simple tactile arrays could discriminate object-surface characteristics such as ridge, edge, pit, crack. Strain-gauge technology, semiconductor arrays, and conductive polymers are being brought into service for force, torque, slip, and simple pattern sensing. And a number of ingenious proximity-sensing systems (optical, ultrasound, etc.) are being looked at as perceivers intermediate to touch and vision.

Commercially available touch sensing has not yet passed the limit-switch, force-sensing options which have been provided for some time by the manipulator manufacturers. However, it seems to be common-place now for users to purchase conventional robots and then modify them for specific applications.

Advanced industrial robots for assembly applications will prob-ably require vision and an anthropomorphic hand equipped with finger-mounted touch sensors to measure the force of the grip. These may be microswitches that sense pressure and make the hand close around the part. Another approach is to cover the hand with rubberized materials whose electrical resistance changes with varying pressure. Several sen-sors to operate a grinder for deburring castings and other industrial robots under development combine touch and adaptive control to grind away welding beads by feeling the actual contour of the weld.

Optical Shaft Encoders

Control of a robot requires knowledge of the position and velocity of each joint. There are a number of ways to get this information. Among the more recent and more accurate techniques are those which involve a particular kind of transducer, a calibrated rotary shaft called an *optical shaft encoder*, and special purpose hardware to interface the encoder to the computer (Snyder & Schott, 1980). Optical encoders measure

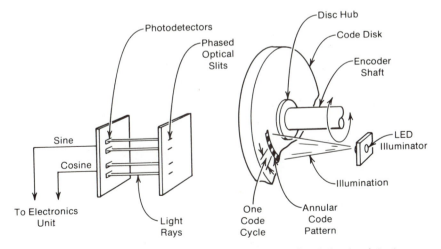

FIGURE 5-2.   An optical encoder measures shaft rotation by detect-ing the light that passes through a rotating code disk.

shaft rotation by detecting the light that passes through a rotating code disc and through a fixed slit or group of slits, as shown in Figure 5-2.

There are two basic types of optical shaft encoders, *absolute* and *incremental*. The incremental shaft encoder can provide a high degree of resolution at a relatively low cost. However, in some applications the absolute optical shaft encoder is easier to use and can avoid certain problems in information transmission.

Incremental encoders have a single code track on the disc, and angular position is determined by counting the pulses produced by modulated light falling on the photodetectors. Direction is sensed by the use of quadrature signals provided by appropriate phasing of the slits.

The absolute encoder consists of a circular glass disc imprinted with rows of broken concentric arcs. A light source is assigned to each row with a corresponding detector on the opposite side of the disc. The arcs and sensors are arranged so that, as the light shines through the disc, the position of the shaft can be uniquely identified by the pattern of activated sensors. Since the absolute encoder assigns to every location a unique coded number, it may be read directly by the computer, and much of the interfacing hardware for an incremental encoder can be eliminated. The Rhino robot uses optical shaft encoders on each joint, as may be seen in Figure 4-8.

6

Vision Sensors and Systems

Vision is an important human sensory power. The eyes sense color since they are panchromatic, and sense motion since the retina provides a lively response to events as they occur. With a hundred million rod and cone cells in the retina, the eye's receptor array is comparable to a grid of pointlike sensors, ten thousand to a side. Therefore, the eye is capable of great detail.

Vision provides shading of light and dark in lines and contours of infinite variety, which intelligence assembles into surfaces and objects, even when the objects have never been seen before. Viewers easily differentiate between a shadow and a dark object, and between surfaces and marks on the surfaces (Albus, 1981; Ballard & Brown, 1982; Haitt, 1981; Nevatia, 1982).

Can science and engineering ever fully explain, model, and duplicate the process of human vision in machines? As a form of sensory data input—a three-color, two-dimensional stream of cues about the dynamic character of the environment—vision has already been recreated in some limited but not insignificant ways. Visual sensors, linked to computers, pick up patterns of light and dark and turn the scenes thus composed into forms that can be processed. The sensors have been linked experimentally to robot arms and manipulators and enable robots to recognize objects by the silhouettes, as long as the objects contrast sharply with the background and do not touch each other or overlap. Such commercial vision systems are rapidly taking their place in modern factories.

Computer vision is the process of producing useful symbolic descriptions of a visual environment from image data. Since its nature and the processes by which it is developed often depend on the uses to which it will be put, computer vision is best understood as the perceptual component of a larger computational problem-solving system.

As with human vision, visual sensors can be considered to be the most important robotic sensors. Visual sensing transducers are usually TV cameras that scan a scene and convert a raster of reflected light intensity values into analog electrical signals. The signals are generated by opto-electrical devices, such as vidicons and solid-state linear or area arrays, preprocessed in hardware, and fed serially at a rate of 30 to 60 frames per second into a computer. The computer analyzes the data and extracts the required information, such as the presence, identity, stable state, position, and orientation of an object to be manipulated, part integrity and completeness of assembly under inspection, and the like. An intelligent robot with vision can, for example, identify parts from machine perception (i.e., computer analysis), assemble and sort them, inspect and grade them, and take corrective action to maintain uniform quality.

Video Cameras and Signals

Image sensors include vidicon cameras and charge-coupled devices (CCD's). The latter consist of arrays of photosensitive elements on a single IC with a built-in lens to focus the image. Cameras based on these chips are usually smaller and more accurate than vidicon-based systems, but the complexity of the chip adds substantially to the cost. Vision systems may be fixed overhead to observe the robot and work piece, or mounted on the robot itself.

Charge-coupled devices (CCD's) are chips that are a special type of semiconductor LSI circuit. Some have hundreds of thousands of light-sensitive elements called *pixels* (picture elements). When an image is focused on a CCD each pixel provides an electronic signal proportional to the light intensity it receives. The computer analyzes the output of pixels and compares the signals with perfect "images" in its memory. It can make decisions and measure objects to within a few millionths of an inch accuracy.

The conversion process begins by partitioning the image into a rectangular array of pixels. At each pixel, the digitizer measures the average brightness, assigns an integer gray-level that corresponds to that brightness, and stores it in computer memory. The digitized image, therefore, is typically stored as a two-dimensional array of gray-levels.

Each gray-level's location in the array—its row and column position—corresponds to the pixel's location in the image.

The spatial resolution of a digitized image is related to the number of pixels in each image row and column defined by the digitizer. The theoretical resolution of an image refers to the volume of information available. When more pixels are sampled in an image, the image is usually represented with more accuracy, up to the theoretical limit. The human eye detects an image with a resolution of more than 4000 × 4000 pixels; the image quality of standard broadcast television has a resolution on the order of 480 × 320 pixels. The SRI vision module shown in Figure 6-1 provides an array of 128 × 128 pixels.

The phenomenal precision inherent in CCD's is what makes them so attractive to industry. The vacuum-tube cameras that they sometimes replace depend on an electron beam scanned across an image target to create a TV signal. That beam, easily deflected by external fields, isn't sufficiently reliable when accuracy is critical. In addition, the analog video signal must be converted via a video image digitizer to a digital representation for processing by a computer ("Video Signal Input," 1981). By comparison, a CCD is self-scanning, with its precision per-

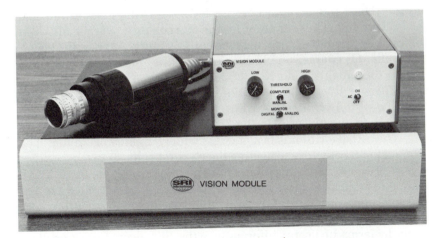

FIGURE 6-1. The SRI vision module for sensing and processing visual images. It can determine the identity, position, and orientation of parts and holes. The vision module consists of three major components: a GE Model TN-2200 solid-state TV camera with 128 × 128 elements; a DEC LSI-11 microcomputer with a 28K word memory, storing a library of the vision subroutines previously developed by SRI; and an interface preprocessor performing general-purpose functions in hardware. *Courtesy of SRI International.*

manently etched into its silicon structure, and it sends out signals in periodic "packets" of information that computers digest easily. These features, in addition to their low power requirement, extraordinary sensitivity to light, long life (no filament to burn out), and ruggedness make CCD's ideally suited for computerized robots.

The Binary Algorithm

A major step in the field of machine vision occurred when Stanford Research Institute developed an algorithm for a simple version of the vision process that provided useful answers in many situations, and was at the same time reliable, efficient, and easy to compute (Agin & Duda, 1975; Reinhold & Vanderburg, 1980).

The inherent difficulty of the vision problem is illustrated by how many simplifying assumptions were necessary to make the SRI algorithm work. To begin with, the algorithm assumes that the picture has been reduced to a binary image. This reduction, which is done either by high-contrast lighting or by some other preprocessing technique, presents the picture to the computer in the following simple form: The picture is stored as an array of dots or pixels which are either on or off (1 or 0). The 1 pixels represent the object while the 0 pixels represent the background. Thus, in order for the algorithms to work at all, useful information about the object must be entirely represented by the object's silhouette. Of course, no color information is understood, and information that might be derived from texture, shading, or three-dimensional perspective is also lost. The algorithm further requires that the object be entirely contained within the field of view of the camera, and if more than one object is in the field of view, they may not overlap.

The SRI algorithm is fast and it is widely applicable, in that it can be operated on any image meeting the above specifications. In addition, it is easy to use, and the information it provides is useful in a wide variety of industrial circumstances. In terms of speed, the simplicity and deterministic nature of the algorithm, together with recent advances in microcomputers, allow operating rates of six images per second or more. In terms of useful information, the SRI algorithm can reliably report on the position and orientation of an object, as well as such fundamental properties as its area, moments of inertia, perimeter, ratio of perimeter to area (a measure of irregularity), maximum radius, and a variety of other powerful geometric invariants. Furthermore, the algorithm will produce these invariants repeatedly and reliably for a wide variety of objects and conditions. Its one principal drawback is the need to reduce information to a binary image, which generally requires high contrast lighting and a carefully structured viewing area.

Probably the most attractive aspect of the SRI algorithm is its ease of use. The key parameters about an object are entered into the computer by a simple process called *training*, whereby the object is shown to the camera in several positions and the computer statistically accumulates all the information it needs. Training can be carried out in a matter of a few minutes by an unskilled operator.

Several U.S. firms offer vision systems for robots using the binary algorithm. One example of such a system is the model VS-100 manufactured by Machine Intelligence Corporation, shown in Figure 6-2. The system includes a Digital Equipment Corporation LSI-11 microcomputer, custom image processing circuitry, interfaces to four types of TV cameras, and a computer interface and software—all packaged in a box about the size of a stereo amplifier.

The VS-100 accepts images from a television camera and determines the type, location, and orientation of objects in the images. This information is then fed to a robot's central computer via a standard interface. The system recognizes objects by comparing features of their

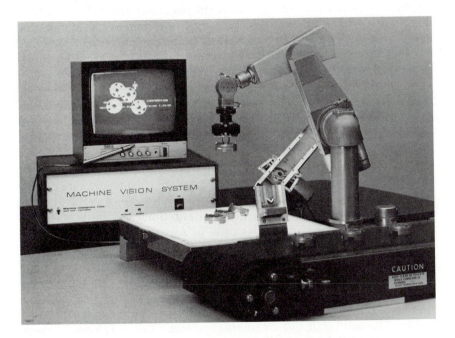

FIGURE 6-2.   The Machine Intelligence Corporation VS-100. A camera mounted on a Unimation Puma robot examines the parts lying on the illuminated table, and a binary picture of the objects is displayed on the screen. *Courtesy of Machine Intelligence Corporation.*

silhouettes to those of exemplars shown to it beforehand during training. As identifying features the VS-100 uses about a dozen geometric properties of an object's silhouette, such as perimeter length, area, and geometric center of gravity. To simplify recognition, system developers chose features independent of the silhouette's location and orientation. They also made the feature set broad enough to identify any possible silhouette. In fact, the full feature set is often unnecessary and is reduced by the operator during system training to speed up recognition.

Pattern Recognition Systems

Computer applications for pattern recognition do not require the quantity or the quality of imagery that the human eye provides. In fact, for many industrial applications such detail and quantity prohibit the storage, identification, and functional retention of images.

The results that any optical recognition technique can achieve depend strictly upon the questions the procedure must answer. The type of pattern recognition employed must correspond to the pattern's physical peculiarities and emphasize the necessary areas of distinction. Requirements other than physical pattern differences can dictate the method utilized. Physical parameters, such as movement, ambient lighting, and orientation, affect the type of recognition system that would be most effective and reliable for a particular application.

An attribute measurement system calculates specific qualities associated with known object images. *Attributes* can be geometric patterns, area, length of perimeter, or length of straight lines. Such systems analyze a given scene for known images with predefined attributes. Attributes are constructed from previously scanned objects and can be rotated to match an object at any given orientation. This technique can be applied with minimal preparation. However, orienting and matching consume significant processing time, and are used more efficiently in applications permitting standardized orientations. Attribute measurement is effective in parts segregating or sorting, parts counting, and flaw detection.

The vision system being investigated at the National Bureau of Standards provides both depth and part-orientation information to the robot control system. The principal components of this vision system are a solid state camera, a structured light source, and a camera interface system. The solid state video camera produces a 128 × 128 raster image. The structured light source is a stroboscopic light which emits a plane of light through a cylindrical lens. The camera and light source are mounted on the wrist of the robot. The camera is oriented to position the columns of the image perpendicular to the flash plane so that each

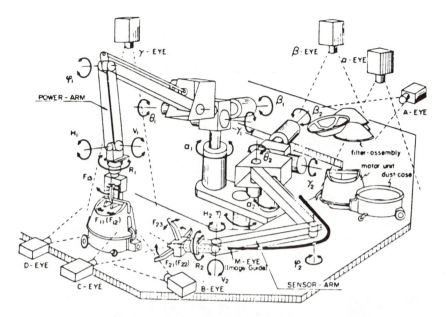

FIGURE 6-3. The Hitachi experimental robot vision system using
visual and tactile sensors.

column of the image has, at most, one intersection with the plane of
light. By mounting the camera and light source at a fixed angle, it is
possible to use triangulation to compute the distance from the robot to
each point in the raster image. When an object is in front of the robot,
the plane of light forms a line segment image. The height of pixels in
the line segment image represents the distance to the camera, while
the shape of the line segment indicates the part orientation.

The final component of the hardware is a camera interface. The
interface, which is attached to an 8-bit microprocessor, provides both
control functions and data reduction. Processing and feature extraction
algorithms are used to obtain the information from the mass of data in
an image. The operations involved may include pattern location and
centering, signal smoothing, equalization or normalization of patterns,
and feature extraction. A sketch of an experimental robot vision system
is shown in Figure 6-3.

The Bin of Parts Problem

In industry, systems linking image sensors, visual processors, and robot
arms are limited to simple, repetitive processes. So far, the machines

can easily be overwhelmed. For example, vision-guided robot arms developed by the automobile industry can position themselves with respect to parts moving on a conveyor, then pick up a part and position it for a later operation. But variations in the position of the objects, as would be the case in a bin of jumbled parts, still confound such image-processing systems.

This bin-of-parts problem is classic. It entails the identification of objects in three dimensions, a more difficult task than identifying parts that present themselves as silhouettes against a surface. The challenges include confusing backgrounds, difficulty in segmenting parts, and the variety of parts that may exist. It makes a big difference whether parts are dark, like the inside of a box, or shiny.

The bin-of-parts problem really represents a threefold challenge in computer vision: to isolate a part amid a jumbled background of the same and different parts; to pick up and orient the part with a robot manipulator; and to set the part into a machine for further operation.

FIGURE 6-4. The University of Rhode Island Mark IV robot with a parallel jaw gripper, alternately picking connection rods and yokes out of a bin using a vision system. *Courtesy of the Robotic Research Laboratories, University of Rhode Island.*

A research team at the University of Rhode Island is working to solve the problem in a general way, not just for a specific type of part or task. They have applied their computer-vision techniques to connecting rods, automobile cylinders, conduit junction boxes, and other objects. They have developed a set of vision algorithms and matched them to work on various classes of objects, and are continuing to look for more powerful, more general algorithms and practical applications (Haitt, 1981). A photo of the University of Rhode Island system is shown in Figure 6-4.

An advanced two-armed robot under development at Hitachi assembles vacuum cleaners using both tactile and visual robot sensors. This system, shown in Figure 6-3, does not use pallets of fixtures for parts location; multiple cameras are used to control the arm(s) locations relative to the vacuum cleaner parts. Template matching, using salient parts features, is used to provide spatial cues to the cameras. The computer control is distributed and hierarchical (Kruger & Thompson, 1981).

Fiber Optics

Fiber optics offers a practical, economical means of providing a vision system for a robot. A fiber optic scanner is made up of a large number of individual glass fibers (a bundle) typically .003 inch (0.08 mm) in diameter, enclosed in a sheath of protective material. Apertures at either end are ground and polished to achieve maximum transmission efficiency at the interface.

A fiber bundle can be coherent or noncoherent. Coherent fiber optics are used for carrying optical images as in a flexible fiberscope. Noncoherent fiber optics are bundled fibers which have not been oriented in any special manner. They are used primarily in industrial applications as light conduits and as such are very efficient and less expensive.

Fiber optics with existing electronic controls can be used in many types of applications at a significant reduction in engineering, design, and cost. Fiber optics are immune to electrical pickup; they also withstand shock and vibration, and are rugged and easy to apply (Wilson, 1981). They are well suited for use in robotic applications, as robots often operate in hazardous locations and at high temperatures.

Vision System Applications

As noted earlier, the Machine Intelligence Corporation VS-100 employs the SRI binary algorithm, and can join a robot for tasks like recognizing

FIGURE 6-5.   The automation vehicle developed by SRI is propelled by electric motors and carries a television camera and optical range finder in the moveable "head." The sensors are the bump detector, the TV camera, and the range finder. *Courtesy of SRI International.*

and picking up randomly oriented parts on an illuminated conveyor belt. Automatix Inc. of Burlington, Massachusetts manufactures Autovision II which is capable of processing 16 gray levels (Iverson, 1982).

On the basis of work done at SRI, General Motors has developed a system known as Consight to enable a robot to identify and pick up variously shaped parts randomly placed on a conveyor. This system

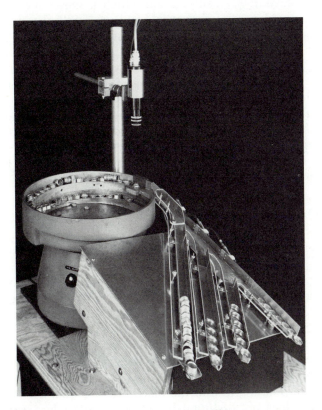

FIGURE 6-6.   A bowl feeder with an eye. This system uses an ordinary bowl feeder combined with a vision capability for the programmable feeding of small parts. Bowl feeders, aided by mechanical wipers, are typically employed to present small parts in a specified stable state and orientation to automatic assembly machines. The mechanical wipers have been replaced with a SRI vision module and compressed air hoses, and added computer-controlled gates for part sorting. The resulting system, called the "eyebowl," is entirely programmable in that it can be easily trained to handle an assortment of different parts. Furthermore, the system can simultaneously inspect the parts and reject those that appear defective. *Courtesy of SRI International.*

uses two tungsten bulbs and a cylindrical lens to project a line of light across the moving conveyor belt, which is viewed from above by a camera. When a part passes under the light it shifts the thin line projected by the lens. The camera, which sees the belt as brightness and the part (which has diffused the line of light) as darkness, is able to establish the shape of the part from the black areas with the aid of a minicomputer.

SRI has developed an automation vehicle using a television system for obstacle avoidance, as shown in Figure 6-5. SRI also developed a bowl feeder with vision, as shown in Figure 6-6.

Space exploration, deep-sea mining, prosthetics, and manufacturing are just a few of the potential applications of sensory-based robotics, a field that provides a growing intellectual challenge to researchers developing the specialized equipment and methodologies required in these and other application areas. Although other sensing schemes—force, tactile, and active/passive compliance—have important roles to play, computer vision appears to offer the richest source of sensory information for intelligent robotic manipulation in a wide range of environments (Jarvis, 1982).

Future Developments

Future robots may be equipped with binocular vision systems, enabling them to see three-dimensionally, much as people do. Such a system is now being developed at the Artificial Intelligence Laboratory of the Massachusetts Institute of Technology. The goal is a system capable of computing the distance to any point in its field of view. The heart of the system is a stereo matcher, a device that combines a pair of binocular TV images to produce a table of distances to the points in the image, known as a *depth map*. The stereo matcher compiles the depth map by first determining the distances to edges of objects in the field of view and then interpolating to find the distances of points between edges. The matcher does not distinguish between "real" edges, such as the rim of a cup, and artifacts such as the edge of a shadow cast by the cup. The matcher under development will be able to produce a depth map from two 100 × 1000 pixel images in from one to four seconds, depending on the degree of depth resolution required.

Solid state cameras with higher resolution and precision are needed. In addition, improved quality of pixels resulting in fewer defective elements and more uniform sensitivity would be a valuable improvement. Also, improved lens design, illumination control, and algorithm design will add to the usefulness of vision systems.

7

Computers and Artificial Intelligence

Early robot manipulators developed in the 1960's operated in record/ playback mode controlled by simple wire-logic hardware. However, in recent years robots have been controlled by minicomputers and micro-computers. The software in such computers makes practical a number of control capabilities that would be prohibitively expensive to wire in. For example, a computer-controlled robot manipulator can record the position of its hand in terms of displacements along the axes of any convenient reference frame. It can then play back that recorded position in a different reference frame to perform a variety of useful activities such as tracking objects on conveyors, tracking objects visually, use of the same program for objects of various shapes, and automatic or semi-automatic calibration for a new work station. Hand positions recorded in this format can also be modified conveniently by the trainer. Fur-thermore, the basic capabilities of a computer-controlled robot can be augmented or adapted to the requirements of a new class of tasks by simply loading a new program, without the need to redesign and rebuild the robot's complex control electronics.

With the availability of the computer, the operator can program the operation of the robot and store it for later use. The operator is able to program the robot using a "teach" or on-line mode, running the robot through the desired movements and then recording the moves for later reproduction or playback. Alternately, the operator can develop a pro-gram off-line and then enter it into the control computer for operational use.

Many robots have been equipped for on-line programming. Whether the robot is designed for point-to-point, coordinated-axis, or controlled-path operation, it has been necessary for the operator to teach the arm by guiding it through its assigned tasks. Generally, this teaching or programming task is divided into three subtasks. The first task is the identification of the coordinates of points at which some action is to occur. Depending on the robot's design, these stored coordinates may represent individual axis joint coordinates or rectangular global coordinates. Frequently, these points are grouped into meaningful paths of motion.

The second task involves stating the functions to be performed at a point or along a path. The functions are chosen from the manufacturer's list of available functions for the robot arm. Typical point functions are tool manipulation and time delay. Typical path functions are velocity of movement, searching and tracking.

The third task is defining the logic of the operation or cycle. This involves determining what path the arm should take under specified conditions. Conditions can be represented by the status of external signal or internal flags, or by the magnitude of internal or external variables. An on-line teaching procedure is satisfactory for many applications, but it becomes very tedious when hundreds of points must be individually programmed. For example, in aerospace there are hundreds of holes that must be drilled in sheet metal parts and many rivets that must be properly placed (Tarvin, 1981). It is inefficient for an operator to manually program these points using current on-line teaching techniques. Instead, an off-line approach to programming, wherein data base information can be used, offers many advantages.

Off-Line Programming of a Control Computer

Off-line programming can be defined as the task of programming the robot through the use of remotely generated point coordinated data, function data, and cycle logic. It eliminates the need for each point to be taught with standard lead-through methods. When off-line programming is used, the robot remains in operation while a new program is being generated. Through off-line programming, the robot becomes more closely integrated into the total manufacturing system, since information from the CAC/CAM data base is shared with other elements of the system. This is a major step toward the totally automated factory.

An example of a robot system capable of on-line or off-line programming is shown in Figure 7-1. The IBM RS-1 functions within a

FIGURE 7-1.   The IBM RS 1 may be operated either on-line using a teach pendant, or off-line using a computer console. Instructions can be given to the RS 1 through either the system's keyboard/display or the pendant shown in the left hand of the operator. The system's arm, at center, is holding a small cube in its gripper. The arm can move in six directions at speeds up to 40 inches per second and can perform a variety of precision assembly, parts insertion, and other intricate manufacturing operations. The RS 1 currently installed at several IBM manufacturing facilities, operates under control of a powerful and easily used programming language—AML (A Manufacturing Language)—developed by IBM specifically for robotic applications. AML permits the RS 1 to respond moment-by-moment to changes in its work environment. For example, it can automatically realign a misfed part in order to complete a task. *Courtesy of IBM Corporation.*

rectangular frame 6 × 4 × 3 feet. The mobile arm is hydraulically powered. The system uses AML (A Manufacturing Language), IBM's unique robotics language, which works with a modified IBM Series/1 small computer. The computer system includes disk and diskette drives, a 120-character-per-second printer, and a keyboard display. In addition to system control, the computer also provides standard data processing functions, such as records keeping, calculation, and report generation. Programmable and diagnostic safety features, such as the ability to monitor the arm's position, are also provided under computer control. An application development program written in AML, which features powerful English-language commands such as *grasp* and *transport,* is available for start-up and initial feasibility studies. A hand-held, push-button control device, called a *pendant,* can be used to move the RS-1's arm through the motions necessary to perform its work tasks. These motions are set in the computer's memory at the operator's command by pressing a button on the pendant.

A benefit of off-line programming is that the job of specifying how the robot operates is taken off of the shop floor and put in the hands of an applications or manufacturing engineer who understands how the robot operates and how the data base information fits into the operations. Thus, off-line programming facilitates decision making and control at a higher level. However, while off-line programming offers many advantages, the concept has its own set of unique concerns. These include:

1. A need to adjust for inaccuracies associated with the arm.
2. A way of aligning various coordinated reference frames.
3. A procedure for interfacing the data base information with the cycle logic and function information.
4. A verification procedure.

The first is particularly important when dealing with large arms designed to lift heavy loads (i.e., more than approximately ten kilograms), but all four concerns must be addressed.

When performing on-line programming, the operator is primarily concerned with the repeatability of the arm, that is, the expected variance in position each time the arm returns to a specified (programmed) point in space. With off-line programming, accuracy of the arm—its ability to go to a point that is commanded rather than physically taught—is of primary importance.

Off-Line Programming Languages

Off-line programming defines and documents robot instructions better than manual teaching. Advanced robot languages permit the user to specify sequences of movements and operations at a keyboard terminal connected to the computer. Software aids available with many of the languages make programming faster and more accurate; for example, subroutines describing frequently repeated steps may be quickly coupled together to build a complex program. Programs are readily modified at the keyboard by using editing routines and symbolic data references. Off-line programming languages provide for more flexible use of sensor data and adaptive control.

Off-line programming languages are classified broadly as either explicit or implicit. Explicit types such as VAL, EMILY, SIGLA, and WAVE permit detailed control over the manipulator actions with direct commands such as as *open, move,* and *pick.* Implicit languages such as AL, ROBOT APT, AUTOPASS, RAPT, and MAL are those in which the user describes the tasks to be performed rather than detailed robot motions. With these languages, the user enters general task commands such as "place crosspiece on bracket such that crosspiece hole is aligned with bracket hole." The program then selects grip points, approach paths, and the motion required to assemble the parts. These implicit languages are primarily experimental.

The logic of the operation of the robot uses a set of instructions of a defined high-level language. VAL, used with the Unimation Puma arm, consists of one-word commands and information the commands operate upon. Examples of commands are *open, move, speed,* and *depart.*

AL, a language developed at Stanford University, makes use of the data base called a world model. AL is a highly structured language with features similar to ALGOL. The language has a number of unique features. It has the ability to work in many different coordinate systems, and will also provide for the simultaneous control of more than one robot either asynchronously or cooperatively. In addition, the language will allow a task to be specified at several different levels of detail ranging from very explicit and detailed manipulator control programs to programs written in terms of high level assembly operators which the system will then translate into manipulator control programs. Language statements consist of motion commands such as *move arm to (position).*

An efficient robot programming language should be easy to learn, type, read, and develop. Also, it is useful if the programmer can make up his own readable, meaningful names for positions, tools, directions, forces, speeds, work pieces, and so on.

Artificial Intelligence

As the automated features of off-line programming become more refined, the language visible to the user is expected to gradually shrink to a minimum. Future robots using these languages will require less input from the user and will rely more heavily on artificial intelligence and sensory input to determine required motions. Such adaptive robots will observe their surroundings by extracting data from appropriate sensors. These data will then be processed by a computer, which will then provide manipulator command instructions. For example, the robot could determine how to assemble different parts based on their shapes. Such an intelligent sensing robot could adjust automatically to virtually any condition.

The extensive use of computers for problem solving has led to a discussion of whether computers may be said to possess intelligence and engage in the process of thinking. This subject of discussion and research is called *artificial intelligence* (AI). Other fields within AI include problem solving using computers, computers that learn, and neurological information-processing models. One definition of *artificial intelligence* is the characteristic of a computer system capable of thinking, reasoning, and learning (functions normally associated with human intelligence) (Dorf, 1981).

A considerable range of human behavior can be explained in terms of information processing theories. Within the theories of artificial intelligence, comparisons can be made between men and machines in that range of activities we describe as "thinking." If use of this term raises objections, then "ability to process information" or some similar term can be used; but it must be admitted that there exists some field of behavior in which men and machines coexist and in which they can be compared. It has been noted that one often regards an action as "intelligent" until he understands it—in explaining the action, it often becomes routine and mechanistic rather than intelligent.

Intelligence may be seen as a summation of several faculties. These include:

- Data capture ability
- Data storage capability
- Processing speed
- Software flexibility
- Software efficiency
- Software range

The central goals of AI are to make computers more useful and to understand the principles which make intelligence possible (Winston, 1977). Using AI technology, a computer can be used for inspection and robot control ("Artificial Intelligence," 1977). One type of system being developed is called an *expert system*, for it models in a computer the knowledge possessed and exercised by an expert in a narrow, confined discipline or pursuit. Since the expert cannot be housed within the mainframe, AI researchers attempt to draw out of the expert the knowledge he or she has acquired from years of experience—"the knowledge of good judgment, the art of good guessing." This experiential knowledge, which AI people call heuristic knowledge, is then combined with textbook knowledge (Davis, 1982).

An expert system consists of two parts: the knowledge base that contains the facts and heuristics of a particular discipline—manufacturing of a part, for example; and an inference procedure with which to reason about the knowledge base. The latter fragment, without the knowledge base is, in effect, an inference engine. The application of AI principles will be evidenced in vision and inspection systems during the next few years.

Japan's Ministry of International Trade and Industry (MITI) is developing a proposal for a major project to develop intelligent robots capable of operating in harsh environmental conditions—including high temperature and perhaps ionizing radiation—for disaster relief and equipment maintenance and supervision. The robots would have decisionmaking capabilities, rather than requiring programmed control for every action, and would, as far as possible, be universal types suitable for multiple applications. Planning will not include commercial production, since the project is meant only to develop new robot technology, but participating firms will benefit from technology "fallout" such as innovations in three-dimensional vision and tactile, ultrasonic, aural, and gas sensors. MITI hopes to get the project started in 1983 and expects to spend between $80 million and $160 million over a ten-year period ("Smart, Rugged Robots . . . ," 1982).

8

Applications of Robots

Robots have rapidly evolved from theory to application over the last decade, primarily due to the need for improved productivity and quality. Increased productivity is the most important benefit of industrial robot application today. U.S. manufacturers will adopt robots which can work around the clock, seven days a week, at a fraction of the cost of a nonautomated production line. Consistent quality resulting from exact repetition of programmed functions is another important benefit. Since robots are not subject to boredom, carelessness, fatigue, or emotion they are ideally suited for repetitive, tedious industrial operations. In addition, robots also effect material savings through scrap reduction and lower reject rates.

One of the key features of robots is their versatility. A programmable robot used in conjunction with a variety of end effectors can be programmed to perform a specific task, then later reprogrammed and refitted to adapt to process or production line variations or changes.

Selection of Robot Applications

The robot offers an excellent means of utilizing high technology to make a given manufacturing operation more profitable and competitive. However, robotic technology is a relative newcomer to the industrial scene, and the prospective buyer of robot technology who is accustomed to buying more conventional items will find robot application a highly complex subject.

With the selection of an application, the buyer must define the robot's function. A clear definition of the task the robot will be required to perform is needed in order to determine if one robot will suffice and the degree of sophistication that will be necessary in terms of control and function. It is also important to determine if the function will remain the same during the estimated life of the robot; if the task will change, it will be less expensive, in the long run, to purchase a programmable robot.

The selection of the robot application is relatively complex, since it requires a clear understanding of robot capabilities. Many people identify what they feel are good robotic tasks only to find later that the cycle speed or articulations required cannot be achieved with a robot. Likewise, good applications often go unnoticed due to a lack of knowledge of the full range of available robot skills and abilities (Ottinger, 1981).

One important fact that should not be overlooked is that the robot may accomplish a task differently than a human. For example, in order to perform work on a fixtured part, an operator may have to pick up and set down a series of hand tools. This does not add value to the part—only the application of the tool to the part adds value. A robot may be able to perform the same task more efficiently since it can pick up the part, hold it in its hand, and take it to the tools, each of which is in a fixed position.

In evaluating a potential application it is necessary to account for performance requirements such as cycle times, parts tolerances, and layout requirements. In addition, the product characteristics and process modifications must be considered. It is also necessary to review the requirements for payload, reach, stroke, memory, complex programming, controls flexibility, human factors, maintenance skills, and cost (Ottinger, 1982).

Industrial Applications

Robots are used today primarily for welding, machine loading, and foundry activities (Robot Institute of America, 1982). As shown in Table 8-1, the number of robots used for welding accounts for 36% of the total robots in the U.S., primarily due to the fact that the automotive industry—the major user of robots, as shown in Table 8-2—has placed heavy emphasis on robotic spot welding. The sharp visibility given to the automotive industry's robotics applications and its declared intention to even more aggressively increase the installation rate has made that industry a major focus for robot builders. Therefore, the automotive

TABLE 8-1.   Applications of Robots in the United States, 1981

	Number of Robots	Percent of Total
Welding	1500	36%
Machine Loading & Unloading	850	20%
Foundry	840	20%
Painting	540	13%
Assembly	40	1%
Other	400	10%
TOTAL	4170	100%

industry will probably retain its position as the leading user of robots, although the applications picture is likely to change as improvements and innovations in robot technology occur.

In the U.S. metal working industry, machine loading and unloading appears to be the biggest single area of application for robots. Assembly will probably run a close second in the future.

Arc welding by robots is a growing application, but rudimentary vision for seam tracking needs to be developed before its full potential will be realized. Also coming into its own is the use of robots for stamping; the robot's lack of speed has been the deterrent up to now but is expected to improve. Changeover of presses and robots in less than an hour will be an important advantage. Conveyor transfer of parts and their palletizing is a challenging application. Robots are being designed which will pick up parts from a supply conveyor and move them to an operation-serving conveyor.

One manufacturing process that owes much of its current growth to robotics is investment casting. The series of steps in pattern making can now be performed easily and consistently. The assembly robot is getting intense attention in the U.S. and Japan. In assembly, the driving impetus is economics, and the multi-shift capabilities of robots may revolutionize one-shift assembly operations. Die casting, injection

TABLE 8-2.   U.S. Industries Utilizing Robots in 1982 in Rank Order

1. Automobile
2. Electric Machinery
3. Plastic Moulding Products
4. Metal Working
5. Iron and Steel

TABLE 8-3. Robot Applications Characteristics

Application	Load	Speed Required	Movement	Accuracy Required
Welding	Medium	Medium	Contour	High
Painting	Low	Fast	Contour	Low
Machine Loading	Medium to Large	Medium	Point to Point	Medium
Assembly	Low	Fast	Complex	Very High

FIGURE 8-1. A large robot used for automatically moving and depositing rolls of clear plastic film to proper pallet locations. This robot can carry 1250 lbs. with a vertical travel of 100 inches and a horizontal travel of 50 inches. It can revolve up to 360°. *Courtesy of Positech Corporation.*

molding, heat treating, and glass handling are other areas of major robotic interest.

The general characteristics of different robot applications are given in Table 8-3. Obviously, each application necessitates a different robot design, including associated fixtures, peripheral devices, and sensors.

Machine Loading and Unloading.   In industries incorporating heavy parts in their products, machine loading and unloading is a natural application for a robot. A robot used for moving a role of plastic film is shown in Figure 8-1.

In many cases, robots can dramatically reduce the need to invest in parts transfer equipment designed for one particular part or process. Fixed automation equipment can be supplanted by a combination of programmable robots and parts transport equipment—conveyors or pal-

FIGURE 8-2.   A Unimate 4000 Series robot removes a 165 lb., 2000°F billet from a forge press to a draw bench where it is formed into an automobile gasoline tank. This robot can lift up to 450 lbs. (205 kg) and move with a repeatability of ± .08 inch (2 mm). *Courtesy of Unimation Inc.*

lets, for instance. Since one can often produce different product styles by simply reprogramming the same robot, the same manufacturing line can produce different product batches. This brings the benefits of automation to factories whose production volumes would not justify customized "hard" automation (Mangold, 1981).

Robots are excellent for handling hazardous materials such as radioactive fuels and heated parts. The robot shown in Figure 8-2 is moving a billet to a draw bench to be made into a gasoline tank. An example of using a robot to quench a part in cooling water is shown in Figure 8-3. Robots can be used to handle large and small parts for palletizing; the robot shown in Figure 8-4 is capable of moving and palletizing 23 fluorescent tubes.

A simple robot can be utilized to pick and place boxes stored in a carousel structure, as shown in Figure 8-5. This common storage system used in conjunction with a robot can improve throughput impressively. The relatively simple pick and place operation is often pro-

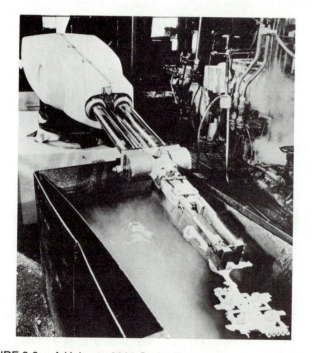

FIGURE 8-3.   A Unimate 2000 Series Robot quenches a part in cooling water after having removed the part from a die-casting machine. This robot can lift up to 300 lb. (136 kg) and has a repeatability of .05 inch (1.3 mm). *Courtesy of Unimation Inc.*

FIGURE 8-4.   A Unimate 2100 Series robot picks up 23 flourescent tubes at a time from a conveyor and palletizes them in tote bins. *Courtesy of Unimation Inc.*

grammed by a lead-through teach operation, as shown in Figure 8-6. The system shown in the figure is capable of moving up to 25 pounds.

The interaction of a loading and unloading device with several processing machines is the core of a flexible manufacturing system. As shown in Figure 8-7, a robot with two grippers can unload a finished part and then load an unfinished part without having to move back to the conveyor.

Painting and Materials Finishing.   Robots can effectively perform a painting operation using a preprogrammed continuous path. In spray painting operations, for example, an experienced painter can teach a programmable robot to paint by going through the motions of painting a workpiece while wearing a teaching device attached to his arm. With the robot in a programmable mode, a microprocessor stores the motions in memory; on playback, the robot, with a conventional spray gun as an end effector, will duplicate the exact movements of the highly skilled painter.

The hostile environment of the automobile paint shop has made

FIGURE 8-5. A vertical-axis MOBOT which interfaces with a carousel structure for automatic storage and retrieval of boxes or totes containing up to 80 pounds of parts or other material. A central computer commands the carousel to present the appropriate bin station to the robot, then commands the robot to lift a tote from the conveyor and place it in the bin, or remove the tote from the bin and lower it to the conveyor. The MOBOT motions are controlled by a programmable controller. Motive power is provided by a controllable speed reversible DC electric motor. Totes are moved on and off the fork by an air cylinder and suction cup gripper. Photoelectric sensors indicate the vertical position of the fork and gripper assembly. *Courtesy of MOBOT Corporation.*

FIGURE 8-6.   A Thermwood Series 7 material handling robot has a continuous path operation. The lead-through teach capability makes it easy to program. The robot has six axes and moves 25 pounds through a vertical stroke of 76 inches and a horizontal stroke of 39 inches with a repeatability of ± .060 inch. *Courtesy of Thermwood Corporation.*

it a prime candidate for the adoption of automation and robot painters, since it removes a human from an undesirable work site. Speed of operation is also increased, and, if properly automated, the robot painter may yield increased quality. A spray painting robot is shown in Figure 8-8. Robots can also be used for materials finishing in processes such as polishing and removing blemishes; for example, Figure 8-9 shows a robot removing burrs from a finished part.

Welding.   As mentioned earlier, the use of robots for welding tasks is the largest application at present, primarily for spot welding in automobile manufacturing. An example of a spot welder is shown in Figure 8-10.

FIGURE 8-7.   A Cincinnati Milacron T3 robot shown loading and unloading a numerical control machine. *Courtesy of Cincinnati Milacron.*

Arc welding is a hot, hazardous, and tediously repetitive job. As
a result of the difficult conditions under which welders work, their
average arc time is only about 30%. In addition, the industry is facing
a shortage of highly skilled welders. A robot welder can increase arc
time, produce more welds of consistent high quality, and avoid the
boring, hazardous elements of the job. The welding robot works continuously without breaks to increase arc time to an average of 70%.
Every weld it makes is precisely controlled for the highest quality. It
is especially useful for welding in cramped or poorly ventilated areas
since it requires none of the OSHA-mandated protection devices needed
by a human welder. And because one welder can easily position and
control more than one robot, productivity per man-hour can be dramatically increased.

An arc welder is shown in Figure 8-11. This robot can achieve
the normal weave pattern of an arc. Once the robot is programmed to
do a certain weld pattern, it continues to reproduce its instructions
independent of the environmental conditions. Smoke, temperature, or

FIGURE 8-8.   The Series Six is a hydraulically powered, jointed-arm, continuous-path robot designed for spray painting. The arm can move 48 inches at a speed up to 30 inches per second. Programming is accomplished utilizing the lead-through teach method. The lightweight composite arm is counterbalanced, allowing easy movement through the desired sequence. The on-line editing system allows for programming the most optimum spray pattern possible. *Courtesy of Thermwood Corporation.*

radiation from the arc are no longer constraints on the welding process. The use of the robot arc welding system results in higher arc-on time, lower costs, and higher quality welds than are available with manual welding.

Increased productivity in robotic arc welding results primarily from the fact that the robot minimizes nonproductive time in the welding cell. If either two weld positioning tables or an indexing device with multiple fixturing is employed in the system, a welding cell operator can be loading and fixturing new parts while the robot is welding in the other location. This increases the flow of parts through the welding system. The result is reduced lead times, higher output, and more efficient production scheduling. The net result is a reduction in the overall input to output time.

In arc welding, heat from an electric arc fuses metals together. In

FIGURE 8-9.   A PUMA robot deburrs an aluminum heat sink with a rotary abrasive brush. This Unimate 500 has five revolute axes using electric motors. It is capable of moving .5 m/second and achieving a repeatability of ± .1 mm (.004 inch). *Courtesy of Unimation Inc.*

gas tungsten arc welding (GTAW), the arc emanates from a nonconsumable tungsten electrode held above the work piece. A blanket of inert gas (usually helium, argon, or a mixture of the two) protects both the tungsten electrode and the weld metal, since exposure to air would quickly oxidize them. People often refer to the GTAW process by its popular nickname, TIG (tungsten inert gas) welding.

In manual TIG welding, the welder first cleans the work piece of all contaminants, such as rust, dirt, oil, grease, and paint. He then strikes an electric arc, perhaps by quickly tapping the electrode on the work. Once the arc is lit, the welder moves his torch in a small circle until

FIGURE 8-10.   Chrysler Corporation uses 63 Unimate robots on this automotive respot welding line. The robots place over 3500 spot welds per car body. *Courtesy of Unimation Inc.*

the heat creates a small pool of molten metal. When the welder gets adequate fusion at one point, he moves the torch slowly along the seam between the parts to be welded, melting their adjoining surfaces, and feeding the welding rod into the pool of molten metal, just ahead of the arc. The welder must painstakingly control his welding speed, the speed he feeds the welding filler, and his welding current. He moves the welding rod and torch smoothly forward, making sure that the hot end of the welding rod and the hot solidified weld are unexposed to contaminating air. With foot controls, the welder adjusts the current to get proper fusion and penetration in the weld. He judges how much current to apply by the size of the molten metal puddle (Neacham, 1981).

Manual TIG welding is a discipline that demands skill and experience. There are not enough TIG welders to meet industry's need for them, and, naturally, the welders are paid highly for their time. The power-generating, chemical, petroleum, and aerospace industries—all of which use high quality TIG welding—would all benefit from the automation of the TIG process.

The four process variables that determine the quality of the weld are: welding current, arc voltage, travel speed, and filler-feeding speed. The changes in these variables, taken over time, define the weld profile

FIGURE 8-11.   An Apprentice Robot automatically arc welds a heavy crane's jib sections. This robot can achieve an average of 70% of possible arc time. It can do this task in half the time it would require manually. *Courtesy of Unimation, Inc.*

for the TIG welding process. To a large extent, welding current determines the quality of the weld. When the current is reduced, the penetration and the width of the weld are reduced. The automatic TIG welder must adjust the current and process energy as well as adjusting to varying seams (Tomizuka, Dornfeld & Purcell, 1980). An example of a seam welder is shown in Figure 8-12.

Assembly.   Assembly of manufactured goods accounts for 53% of total production time. Usually one-third of a manufacturing firm's work force is involved in assembly tasks (Nevins & Whitney, 1978; Jurgen, 1981). Only a small portion of today's assembly processes are automated: about 6% of subassembly and 4% of final assembly are fully mechanized, and another 17% of subassembly and 10% of final assembly may be performed by semiautomatic equipment (McCormick, 1982).

FIGURE 8-12.   The T3 robot performing arc welding along a seam.
Courtesy of Cincinnati Milacron Corporation.

Mid-volume batches of a product and mixed batches of similar models of a product can be assembled profitably by machines that are adaptable and programmable. An adaptable machine is one that can perform an assembly task as it accommodates itself to relative position errors between the parts. Such errors arise from the usual tolerance within which all manufactured parts are allowed to vary, and from a lack of perfectly repeatable performance by the assembly machine itself. It is these errors that cause parts to jam rather than go together smoothly. Adaptability is therefore central to successful assembly.

A programmable assembly machine is one that can be taught, with minor alterations, to perform a new assembly task or that can perform several tasks in sequence (Nevins & Whitney, 1978). The mating of parts involves all the events that occur as parts touch and go together. Such events are governed by the geometry of the parts, particularly the amount of clearance between them after assembly, by the degree to which they are misaligned laterally and angularly when they first touch, and by the influence of contact and frictional forces between them as they slide together. The difficulty of mating parts is illustrated by the fact that typical tolerance requirements are specified in millionths of an inch.

The Draper Laboratory of MIT recently built and operated the adaptable programmable assembly system shown in Figure 8-13, which consists of a computer-controlled, industrial assembly robot with four degrees of freedom. The wrist of the robot contains a passive-compliance insertion device. The system has been taught to assemble a Ford automobile alternator using six interchangeable tools. The alternator has 17 parts, all of which can be inserted from one direction with the aid of two assembly fixtures. This experiment demonstrated the fea-

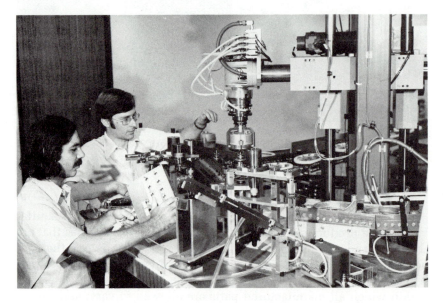

FIGURE 8-13.   Two research engineers at Draper Laboratory are shown teaching an assembly robot one of the operations of assembling an automobile alternator. The robot has four degrees of freedom and can change tools. It is controlled by a Data General NOVA 2 mini-computer using a control language developed at Draper. The robot takes two minutes and 42 seconds to assemble the 17 parts of the alternator and uses three grippers, two screwdrivers, and a nutrunner in the process. The assembly is greatly aided by a passive multi-axis compliance in the robot's wrist called a Remote Center Compliance. *Courtesy of The Charles Stark Draper Laboratory Inc.*

sibility of assembling a stack product with a robot capable of moving along only four axes. However, the alternator may represent a special case—most stack products have some parts that arrive from eccentric directions, which means that a single four-axis robot could not assemble the entire product.

The alternator is a good experimental case because it is a stack product and its assembly calls for several tight-clearance peg-hole insertions and screwdriving operations. The Draper experiment showed that the compliant wrist is an extremely valuable device. Since the roughness of the outer surfaces of the parts makes them difficult to hold in their feeder tracks or on the assembly fixtures in precisely known and repeatable positions and orientations, position errors of one millimeter were typical and unavoidable. The compliant wrist overcomes such errors; it can assemble a complete alternator, picking up parts of various sizes, placing them on fixtures, and inserting even extremely close-tolerance parts successfully. The flexibility of the device also greatly reduces the time required for adjusting the feeders and fixtures. The complete assembly operation takes two minutes, 42 seconds. Improved engineering of the tools and fixtures, together with a simple design change, could reduce the time to about one minute. The biggest saving would be in reducing the time needed for changing tools, which now takes about 30 percent of the total. One route to this end would be to assemble several alternators at once, so that the robot could carry out the same operation on several units before changing to the next tool.

For years, robots have been assembling electronic watches in Japan, typewriter keyboards in Italy, appliance parts in Europe, automobile bodies in the U.S. and abroad, and many other subassemblies all over the world. Robots are also be used to assemble parts for production aircraft and for floppy disk drives.

Various studies and plans are underway in the U.S., Japan, and elsewhere to develop the world's first fully automated plants for heavy manufacturing and batch-lot assembly of small parts. By 1984 the Japanese plan to have in operation an unmanned metal-working plant producing everything from hydraulic pumps to heavy-duty transmissions for vehicles. Every operation from casting to final inspection will be handled by numerically controlled tooling machines, industrial robots, and other flexible automation systems. The plant will be supervised by a handful of engineers and technicians.

One example of a robot assembly operation is the six-robot system used by Sony to assemble a tape recorder mechanism involving 48 parts. Another example of an assembly operation using a robot is shown in Figure 8-14.

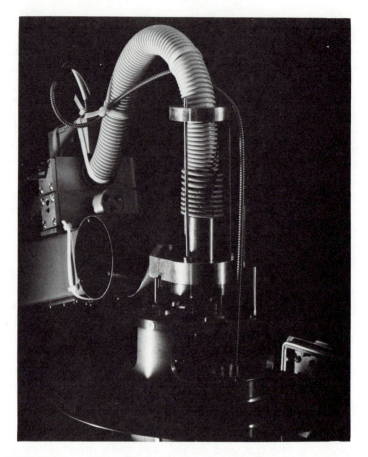

FIGURE 8-14.   This robot is used in the production of head-disk assemblies for computer disk drives where tolerances are a few millionths
of an inch. This photo shows the insertion of a disk on a shaft. *Courtesy
of Storage Technology Corporation.*

A robot is ideally suited to light assembly tasks that are capable
of good repeatability and accuracy over long operating times (Thompson, 1981). An example of an assembly robot is shown in Figure 8-15.

As part of the assembly process, testing of units and subassemblies
is required. The manufacturing process begins and ends with testing—
from incoming inspection to subsystem testing to the out-the-door final
system test. The goal of any manufacturing operation is to produce and
ship products that the customer will accept and pay for, and the ability
to ship a working product depends on the ability to test that product

FIGURE 8-15.   The P300H articulated arm is useful for fast, precise, light assembly work. This robot can move 5 kg (11 lb.) with a repeatability of ± .1 mm (.004 inch). *Courtesy of Par Systems, GCA Corporation.*

in its various stages of manufacture. An example of a robot-assisted assembly test operation is shown in Figure 8-16.

Mobile Robots

Mobile computer-controlled vehicles and mobile robots are candidates for applications that include mining, farming and undersea use, as well as in the factory. A driverless mobile vehicle used in Japan for factory automation is shown in Figure 2-15. This vehicle is guided by wires and can be loaded and unloaded by robots or have a robot mounted on it.

A large, automatically controlled vehicle with manipulators at-

FIGURE 8-16. Visually aided robot arms are used at Texas Instruments to perform pick-and-place operations at a test location on an assembly line for calculators. *Courtesy of Texas Instruments.*

tached could be used for farm operations such as plowing, seeding, fertilizing, spraying, irrigating, and harvesting. These chores can be done by robot devices the way irrigating is done today with center pivots. High-value, speciality crops such as strawberries, houseplants, flowers for the florist industry, Christmas trees, vegetables, and other crops would be well-suited for farming by machines. Twenty-four-hour operation would be possible, especially when weather conditions and crop maturity demand immediate harvesting.

Mobile robots can be used for performing nighttime security tasks for factories, warehouses, and museums (Bulkeley, 1982). Mobile robots moving on wheels will carry sensors to detect intruders, and will patrol the hallways and fences of industrial firms with unflagging attention. Mobile robots such as that shown in Figure 8-17 may be used in large materials handling tasks. Research on mobile, multiple-legged devices continues in the U.S. Underwater remote-controlled vehicles with robot

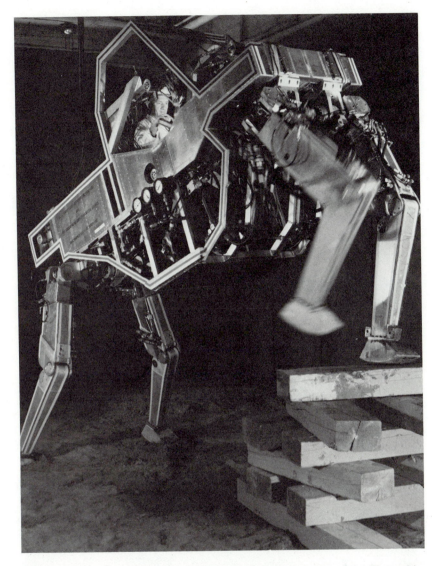

FIGURE 8-17.   This research prototype of a quadruped (four-legged) machine, fabricated by General Electric Company engineers under a U.S. Army contract, was designed to spur development of equipment that will improve the mobility and materials-handling capabilities of the foot soldier under the most severe conditions. By means of an advanced control system, the machine mimics and amplifies the linear movements of its operator. The right front leg of the unit is controlled by the operator's right arm, its left front leg by his left arm, its right rear leg by his right leg, and its left rear leg by his left leg. The research prototype is 11 feet high and weighs 3,000 pounds. *Courtesy of General Electric Company.*

FIGURE 8-18.   One conception of an undersea vehicle used for salvage operations. This vehicle has four articulated arms. *Courtesy of International Submarine Technology Ltd.*

manipulators can be used for underseas mining and salvage operations. The use of an undersea robot vehicle is illustrated by Figure 8-18.

Future Applications

There are large numbers of future applications for robots meeting relatively exotic requirements. For example, a robotic deriveter is being designed for the automatic removal of rivets en masse from naval aircraft, permitting inspection and repair of saltwater-corroded airframe members, rapid implementation of airframe field changes, and nondestructive inspection for cracks around the rivets (with or without rivet removal) (Vranish, 1982). Its development will be significant because it is a flexible, mobile system geared toward disassembly and repair rather than the more traditional assembly and manufacture functions, and because it is the Navy's first major development effort in robotics.

During aircraft maintenance, it is often necessary to remove large sections of the wing skins. Skin panels may exceed 100 square feet (9.3 sq. m.) in area. They vary in size and shape and are attached to the frame by up to 4000 rivets for a single upper wing surface. The rivets also vary in size. The current method of rivet removal—manual drilling and punching of each rivet—is tedious, dangerous work. The repetition and physical and mental effort involved have an adverse effect on morale and quality control. Thus the Navy is developing a robotic arm to remove up to 500 rivets and achieve high accuracy of operation.

An example of the way the Japanese try to promote the use of robots in a highly practical manner can be gained from a research project aimed at using robots in the assembly of panels of various sizes for prefabricated houses. In Japan, houses are built to a number of standard patterns, and some companies, such as Shimizu Construction Co., are now prefabricating panels for subsequent assembly on site. There are two main operations: cutting and assembly of panels, and welding, assembly, and finishing. The system was based on the use of several robot arms operating in tandem (Hartley, 1980).

Robot arms can be utilized for prosthetics and rehabilitative devices. Also, robots can be used to aid nurses in hospitals in the provision of services. For example, Japan has developed a robot linked to a driverless cart that shuttles between a cabinet and a hospital patient's bed. The system can be activated by a voice command or, for patients who can't speak, by whistles and gasps. The arms can find and deliver a newspaper or any other item in the cabinet.

Robots can be used in food processing, and several such proposals have been put forward. One firm has proposed a multifunction "marathon worker" capable of cooking and serving hamburgers (rare, medium, or well done), chicken, fish, french fries, and fruit pies, in addition to serving a wide variety of beverages automatically in response to customer selection. In Japan a robot is being used to make sushi, a raw fish and rice dish. The robot is sold for $6900 and is capable of pressing, punching, and patting into shape about 1200 oblong lumps of rice per hour.

9

Economic Considerations

Many observers have forecast that computer-aided manufacturing will become widespread by the mid-1980's, dramatically increasing per-capita production in the United States and ushering in a whole new era of manufacturing technology. Now, only several years away from this target date, predictions of annual growth rates of only 10 to 30% in the sale of industrial robots and computer-aided design (CAD) manufacturing systems reflect a more sober assessment. Advances in electronics and computer technology over the past decade would seem to justify claims that decreasing component and system costs will make robot systems cost-competitive with human labor and therefore hasten their adoption. What, then, are the factors that have slowed the automation revolution?

Several factors need to be overcome to effect more rapid diffusion of automated manufacturing technologies in the U.S. These include lack of knowledge about robots and flexible manufacturing, inadequate appreciation for their potential, and, especially, the short-term investment horizon held by most American managers.

Part of the problem has been that the potential buyers and users of robot systems have not been as optimistic about the benefits of automated manufacturing as the systems' purveyors have been. Corporate managers are hesitant to commit large sums of money to machinery that has not existed long enough to have proved itself worthy of the investment required. Fearful that technological advances might make their equipment obsolete, they demand one- to two-year payback periods on these systems, compared with eight to ten years for more conventional equipment. A more realistic payback period for automated manufacturing equipment would be extended to two to four years.

Factory floor managers are even less likely to push for the adoption of computerized manufacturing. Most are only partially informed about commercially available systems and how they could be incorporated into the current manufacturing facility. Automation's long-term benefits are considerably less important to these managers than the short-term productivity upon which their performance is usually evaluated. And perhaps most important of all, few factory managers are equipped with the broad background required to manage a combined work force of people and computerized machinery; they are therefore not willing to take part in what they perceive to be their own obsolescence.

Computer-aided manufacturing is thus undergoing a typical learning-curve growth; companies are adopting a "wait and see" attitude and paying close attention to the results of current computer-aided installations.

Lack of standardization has also retarded the acceptance of robots. No group or organization has made moves toward setting any standards, and no single company has dominated the robot industry to the extent that its products have become de facto standards for the rest of the industry, as IBM has done for the computer industry.

It is a basic fact that unemployment in any industry is caused by a decline in its competitiveness; if it fails to adopt the technological advances used by competitors, its employment will decline much more rapidly than if it adopts such advances, even though they may involve some initial displacement of labor. In actuality, many domestic industries have already experienced considerable reductions in work force, and this threatens to continue on an even greater scale if technological lags are not reduced. Even those domestic industries which are gaining competitiveness may now require reductions in man-hour requirements per unit of output of at least 20 to 30 percent. Moreover, many productivity lags are continuing to grow as foreign competitor's efforts to surpass American performance keep intensifying, as may be illustrated by Japanese developments in the steel, automobile, machine tool, and semiconductor industries. Therefore, major improvements in the performance of domestic industry are imperative.

Opportunity is still wide open for U.S. domestic manufacturing to overcome its current lag in this area and thereby achieve major improvements in productivity and cost competitiveness. Although many observers claim that Japanese industry has surpassed the United States in the use of programmable automation systems and robots, such applications still account for only very limited sectors of Japan's manufacturing industry. The systems are even more sparsely used in Western Europe.

Any attempt to delay or slow down the diffusion of programmable automation and robotics can only be justified by identifying and promoting other means of achieving the needed advances in productivity and cost competitiveness within the next five years.

Robots have been, and will continue to be, introduced as direct replacements for individual workers performing manual tasks. But an increasing proportion of robotic applications in the future is likely to derive from continuing development and diffusion of programmable automation systems, which are likely to require corresponding improvement in the capabilities of their robot components. Accordingly, the key issues involved in increasing the contribution of programmable automation systems and robotics to strengthening the international competitiveness of domestic manufacturing industries would center on:

1. The adequacy of the rate of development of their technological capabilities relative to the rate of progress abroad.

2. The adequacy of the rate of their diffusion relative to their capacity to improve productive efficiency and cost competitiveness, and also relative to such diffusion rates among foreign competitors.

3. The relative effects of slower and faster rates of development and diffusion of automated systems on the competitiveness, employment levels, and capital requirements of various domestic industries.

4. The identification of the nature, sources, and relative importance of the influential determinants of change in their rate of development and diffusion.

The formulation of effective approaches to encouraging fuller realization of the constructive potentials offered by programmable automation systems and robotics requires careful exploration of these issues.

Management must recognize that, in order to achieve improved productivity through the introduction of robots and other flexible machinery, the design of the production system must be enhanced. This can only be achieved through improved capital equipment and associated improvement in the design process, renewed emphasis on the structure and operation of the production system, and enhancement of knowledge and skills of labor and technical personnel.

Reasons for Using Robots

Robots are used in factory applications because of several economic and other considerations, as shown in Table 9-1 (Tanner, 1978; Engelberger, 1980). Reduced direct labor cost is the primary consideration in the case of welding, material handling, and machining; however, in painting the reduction or elimination of hazardous, unpleasant work is paramount.

In general, firms that can effectively use robots have a production volume requirement that lies between manual operation and fixed or hard automation as shown in Figure 9-1. Though some robots are installed in development situations or to remove workers from a hostile environment, it is apparent that the prime mover in adoption of industrial robots is productivity. If there is not enough direct and indirect labor savings to justify an industrial robot installation, continued application evaluation is worthless.

The motivations for using robots differ by application, but reduced labor costs and the elimination of dangerous jobs are high priority reasons, as shown in Table 9-2.

The Planning Process

The successful purchase and installation of an industrial robot requires that the entire process be planned and carried out in a logical sequence. Although the basic steps are similar to those followed in the acquisition of any type of automation, robots have unique capabilities and limitations that make it especially important to carefully plan the application and implementation process. The entire process of implementing a robot from the initial planning through the final operation of the robot

TABLE 9-1.   Reasons for Using Robots for Selected Applications

	Reduction in Direct Labor Costs	Increase in Productivity	Hazardous/ Unpleasant Work	Improved Product Quality
Welding	1	2	4	3
Material Handling	1	3	2	4
Machine Loading	2	1	3	4
Spray Painting	3	4	1	2
Assembly	2	1	4	3
Machining	1	2	4	3

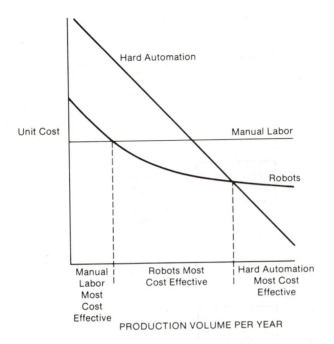

FIGURE 9-1. A comparison of unit costs for manufacturing methods by level of production. Robots are most effective for batch or medium quantity production.

on the production line takes place in four general phases, as shown in Figure 9-2.

The planning phase includes organizing the project team, defining the objectives of the system, and identifying the candidates for the application of a robot. Once the applications are selected, the review

TABLE 9-2. Motivation for Using Robots (In Rank Order)

1 Reduced Labor Cost
2 Elimination of Dangerous Jobs
3 Increased Output Rate
4 Improved Product Quality
5 Increased Product Flexibility
6 Reduced Materials Waste
7 Reduced Labor Turnover

Source: Carnegie-Mellon Robotics Survey, April 1981.

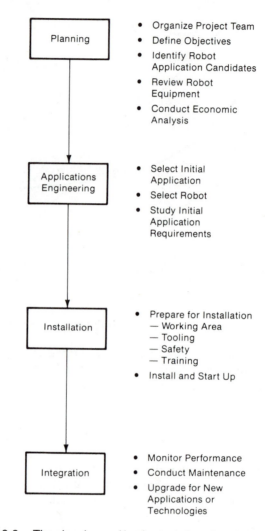

FIGURE 9-2. The planning and implementation of a robot installation.

of robot equipment and an economic analysis of each application candidate can be accomplished. During the applications engineering phase, the candidate applications are studied in detail, and a specific first application and a specific robot are selected (Cousineau, 1981; Owen, 1980). The installation phase covers the activity from the preparatory work performed on the work place through the installation and start-up of the robot. Finally, once the robot has become operational the integration phase is required to insure it continues to perform its job in an effective manner. Activities to be performed during this phase include maintenance, monitoring, and training.

Most robots are required to interact with other machines, transfer lines, or parts from outside their immediate environment. For example, a robot cannot transfer a part until an input signal has been received by the robot that the part has arrived at the initial position. Once the robot has successfully transferred the part to the end position, it must move clear of the conveyor line and signal to the line that the next part can be sent to the initial position. It is in this area of interaction that external sensing capabilities play an important role. Programming of the computer controller and developing suitable interfaces with associated equipment is important for high quality robot applications.

Significant opportunities exist for the emergence of new industries in applications engineering, software development, and systems integration. The review of potential candidates for robot applications requires significant understanding of the manufacturing work place as well as of the capabilities and limitations of robots. The buyer of robot systems is interested in the application, not the robot itself, and it is in the area of applications engineering that the rapid introduction of new robots will be facilitated. Interfacing programs and sensors and software control programs need to be developed in order to enable robots and their associated machines to operate more rapidly and more effectively. Finally, opportunities exist in the area of integrating total flexible manufacturing systems—robots, machine tools, parts handling, inspection, and other associated manufacturing processes. Currently there is a greater need for systems integration companies than for new manufacturers. A common criticism of the industry is that there is a surplus of arms, manipulators, and controllers, but a shortage of proven effective applications. The applications are more complex than the robot manipulators themselves and require extensive systems integration. A few industrial firms are capable of this activity themselves, but most require assistance by systems integration houses.

Costs of a Robot Installation

Robot cost varies with design and function complexity and can be as low as $20,000 or as high as $200,000. For example, a welding robot installation costs approximately $160,000, while a typical material handling robot costs only $60,000. Average cost of a robot installation is in the neighborhood of $80,000 and divides as follows: the robot, 55%; accessories, 30%; and installation, 15%. A typical robot for a mid-cost system is shown in Figure 9-3.

FIGURE 9-3.   The Maker robot system is a five-axis system with a telescoping joint. This arm can lift 5 lbs. and move up to 55 inches per second with a repeatability of .004 inch. This robot can be used for assembly applications as well as operations, material handling, machine loading and unloading, testing and inspection, and manufacturing R&D efforts. *Courtesy of United States Robots.*

Although the initial cost of a robot installation is considerably higher than that of hiring a new employee, that differential is more than offset by the average cost of robot operation as compared to the average remuneration of a skilled worker. For example, labor costs including benefits in the automotive industry were about $8 per hour in 1972 and increased to about $17 per hour in 1982. By comparison, the average cost of operating an industrial robot is $5.00 an hour, and its costs do not increase over time as do those of labor. In addition, employees must undergo considerable retraining when switched from one job to another, while robots can be reprogrammed with little difficulty to accommodate production line changes.

Economic Analysis for Robot Systems

A critical step in planning for the application of a robot is to develop initial cost estimates and perform an economic analysis. Although non-

TABLE 9-3.   Costs and Savings with a Typical Robot Installation

Costs		Savings	
Purchase Price (P)	$60,000	Direct & Indirect Labor (L)	$20/hour
Accessories & Equipment (A)	$30,000	Material (M)	$1/hour
Engineering & Installation (I)	$20,000	Tax Credit (C) (@ 10%)	$6,000
Maintenance & Operation (O)	$2/hour	Annual Depreciation (D)	$10,000
		(straight line, 6-year life)	

economic factors such as worker safety and hazardous conditions are often cited as the main reasons for adopting a robot system, ultimately a cost justification is required. In this section we will consider the cost justification of a robot installation using three financial analysis methods: payback period, return on investment, and internal rate of return. Table 9-3 provides the associated costs and savings for a typical robot installation along with representative dollar amounts that will be used in illustrating the three methods of financial analysis.

Payback Period.   The payback period is a measure of the time required to recover the initial investment. Therefore, if the payback period is two years, a robot system will return net positive cash flow after that period for the remainder of the life of the system. A simple equation for determining payback period is as follows:

$$\text{Payback Period} = \frac{(P + A + I) - C}{(L + M - O) \times H \times (1 - TR) + D \times TR}$$

where H = hours of use during the year and TR = the tax rate of the firm. Assume TR = .40 and a one-shift operation so that H = 250 days × 8 hours = 2000 hours. Then, for the typical costs used in this example we have:

$$\text{Payback Period} = \frac{\$110,000 - \$6,000}{(20 + 1 - 2) \times 2000 \times .6 + 10000 \times .4}$$

Then, we find:

$$\text{Payback Period} = \frac{104,000}{26,800} = 3.88 \text{ years}$$

If the same system were used on two shifts, then H = 4000 hours and we have:

$$\text{Payback Period} = \frac{104{,}000}{19 \times 4000 \times .6 + 4000}$$

$$= \frac{104{,}000}{49{,}600} = 2.10 \text{ years}$$

This equation may be adequate for financial analysis when the payback period is short such as one or two years. When the period is longer, we include the time value of money by using discounted cash flows, but let us first determine the return on investment.

Return on Investment.   Once a firm has ascertained its cost of capital for its new investment it can then compare the return on investment with its cost of capital and determine if it meets or exceeds the cost. The return on investment is calculated as follows:

$$\text{Return on Investment} = \frac{S \times 100}{T} \text{ percent}$$

where

$$S = \text{Total annual savings}$$
$$= (L + M + O) \times H - D$$

and

$$T = \text{total investment}$$
$$= P + A + I - C$$

For the typical costs considered previously and a two-shift operation we have:

$$S = 19 \times 4000 - 10{,}000$$
$$= \$64{,}000$$

and

$$T = \$104{,}000$$

Therefore, we have

$$\text{Return on Investment} = \frac{\$64,000}{\$104,000} = 61.5\%$$

This compares favorably with the current cost of capital, which may be assumed to be 20%, and demonstrates the favorable results of the robot installation. This calculation does not account for any tax consequences.

Internal Rate of Return. The calculation of the internal rate of return (IRR) utilizes a calculation that accounts for the time value of savings realized in the future. In the example under consideration, the total initial expenditure (T) is $104,000 for the robot and associated equipment. The annual savings are experienced over a span of six years, assuming the system is run for two shifts (16 hours per day). The annual savings (AS) are:

$$AS = [(L + M - O) \times H - D] \times TR$$

Again assuming a tax rate (TR) of .4, we have

$$AS = [76,000 - 10,000] \times .4$$

$$= \$26,400$$

which occurs for six years. The calculation for IRR is accomplished by calculating the present value (PV) of the stream of future cash income (AS) for a candidate rate (R) using

$$PV = \frac{AS}{R} \left[1 - \frac{1}{(1 + R)^n} \right]$$

where AS = the annual savings received for n years. When the present value (PV) is equal to the total initial investment (T) of $104,000, then the rate (R) is called the internal rate of return. Calculating the IRR for the example may be achieved using a table of discount factors or by means of a calculator or computer. For this case we have

$$IRR = 15.1\%$$

Total Cost Comparisons. It may be useful for a firm to compare the total costs of using a robot system with that of the normal labor procedure over a selected period, perhaps the life of the robot system or the planning horizon of the firm. We will consider the example discussed in the previous sections and assume a comparison period (Y) of five years with a two-shift operation. For the normal labor procedure the total cost (TCL) is:

$$TCL = H \times L \times Y$$
$$= 4000 \text{ hr. } \times \$20 \times 5$$
$$= \$400,000$$

The total cost of the robot system (TCR) may be calculated as:

$$TCR = (P + A + I - C) + [(O - M) \times H - D \times TR] \times Y$$

Then, for the example we have:

$$TCR = 106,000 + [\$1 \times 4000 - \$10,000 \times .4] \times 5$$
$$= \$106,000$$

Therefore, over the selected five-year period the total cost of the labor system is almost four times that of the robot system. Of course, this favorable ratio would deteriorate if maintenance and operating costs were to increase over time. Again, a reliable, easy-to-maintain system is critical to desirable financial results.

Increased Production Output.   Often a firm using a robot is able to achieve a production rate exceeding that obtained with manual labor. This increase in throughput or product output improves the economic performance of the installation since the economic savings are increased by the factor of the output increase. For example, for the representative system considered in the preceding sections, if the production rate can be increased by 50%, the savings achieved will be increased by 50% (Owen, 1980).

Availability of Capital

The availability of capital for the purchase of automated manufacturing equipment is crucial to rapid diffusion of this equipment throughout U.S. industry. The current high interest rates are a significant barrier, while the investment tax credit offered by the federal government is an important incentive. However, additional incentives are needed to aid further development of the automated manufacturing industry as well as obtain more rapid adoption of this equipment by industry. Given the fact that in this decade Japan is focusing on the rapid development and application of robots and automated manufacturing, the United States cannot afford to stand by and watch while international competition improves its cost position and quality of production. In order to maintain a first-rank series of industries which will be internationally competitive, the U.S. must take every action possible to provide financial incentives for the purchase of automated manufacturing equipment. Such incentives would include subsidized interest rates for capital equipment, an investment tax credit, accelerated depreciation of the equipment, and perhaps encouragement for establishment of private firms which would offer leasing arrangements. If robots and automated manufacturing equipment were leased as readily as railroad cars or computers, prospective users would find their costs spread over the life of the equipment. The financing needs of industry would be greatly assisted by leasing opportunities. The advantages of leasing include lower cost and improved cash flow. In addition, the risk of equipment obsolescence is shifted to the lessor.

Japan offers several incentives which encourage its industries to adopt automated systems. The Japan Development Bank started a leasing company that caters to small firms whereby a firm wishing to acquire a $9000 robot may lease it at a monthly cost of only $300. Japanese companies that prefer to purchase rather than lease a robot can get government aid in the form of subsidized loans. In addition, the Japanese tax code provides extra depreciation for robots, enabling firms to depreciate 52.5 percent of the purchase price in the first year.

Total Cost of Automated Manufacturing Equipment

A critical factor to be recognized in purchasing automated manufacturing systems and robots is the fact that the total cost of installing and operating such an installation is many times the cost of the parts and equipment. For example, the use of a robot on an assembly line would typically require two to three times the cost of the robot itself. The engineering, design, installation, and special support equipment required to integrate the robot into an assembly line often will amount to several hundred thousand dollars. Many firms using several robots, parts handlers, computer control equipment, and associated devices find that automated manufacturing lines cost in excess of several million dollars, though the robots and manipulators themselves cost less than a million dollars. The fact that the firm is required to invest extensively in staff salaries, design activities, and consulting results in a diminished incentive to purchase the new equipment. Support services, integration assistance, and financing arrangements are as critical to the total effort as the automation design required.

Reported Results

Ford Motor Company has 430 robots in service nationwide (Lawrence, 1982). Its assembly plant in Milpitas, California, has twelve Unimate 2000 robots used for welding car bodies and engines. These robots cost $75,000 each but accessories and installation raise the unit cost to $120,000. The Milpitas plant engineer estimates that each robot does 1000 spot welds per hour and will pay for itself in five to ten years.

In Japan, Pentel Co. has designed a small Puha robot which sells for $19,000. According to Pentel, the robot has lowered production costs by 45% and narrowed lead time for bringing new products on line by 80% ("Pentel signs its name . . . ," 1982). The Nippon Electric Company has developed an assembly robot that can position a component with a maximum error of 40 millionths of an inch. They calculate that playback robots—systems that continuously repeat a specific set of motions—in 1976 cost 4.2 times the average annual savings; now they go for only 2.2 times the savings. Japanese robot firms are seeking overseas markets, and it is expected that exports of robots will soar by 1985 to 17% of total output; some analysts are suggesting that by 1990 the number could be 20%. Yaskawa Electric Manufacturing Company, Japan's largest builder of arc-welding robots, is counting on exports to help triple sales and perhaps overtake Kawasaki Heavy Industries Ltd. to become Japan's leading robot maker. Yaskawa exported 16% of its units in 1981 and plans to boost overseas sales to about 20% of its projected output of 1,500 robots ("The push for dominance . . . ," 1981).

The Soviet Union has launched a campaign to catch up with the rest of the industrialized world in robot use. The Soviets see robots as a "quick fix" for two debilitating problems: a shortage of manpower and the lowest labor productivity in the industrialized world. Despite a labor force 140 million strong, every major civilian industrial project in the country is short-handed, and the problem is compounded by a high incidence of drunkenness on the job and sheer alienation that contributes to what Westerners regard as an extraordinarily high rate of absenteeism. However, while robots may be one answer to Russia's manpower shortage—which will become worse after 1990, when the work force is projected to stop growing—they also underscore other problems: the lack of skilled technicians to install and service the units, and the oppressive administrative climate under which plant managers operate. According to a report from Russia's State Planning Committee, fully half of the 5,000 robots produced from 1976 to 1980 sat in warehouses at the factories to which they were delivered, gathering dust for extensive periods. Plant managers simply refused to shut down lines to install the robots, since losing production would mean risking the wrath of the Kremlin ("Russian robots . . . ," 1981).

10

The Worker and the Work Place

The collaboration of man and machine has been a fact of Western civilization since the industrial revolution. However, the adjustment of the American worker to his mechanized counterparts has been, and continues to be, a difficult one. Due to displacement of labor as industries convert to automated equipment, many workers see themselves as victims of automation—"caught in the machine," as illustrated by the photo in Figure 10-1.

In fact, one of the unexpected achievements of the Industrial Age is that labor-saving machinery has added rather than subtracted jobs. But can this continue, given the imminent growth in use of robotics and automated manufacturing? This critical question is just beginning to engage the serious attention of industrialists, labor leaders, sociologists, national planners, and members of Congress.

Population trends in the U.S. point to major changes in the next few years as machines replace blue-collar workers on production lines. During the period 1980–1990, the percentage in the age group 18–24 will decline by 15%. As a result of this relative drop in new workers entering the work force, there will result a relative decline in unskilled workers. There will therefore be incentive and opportunity to use robots and automated manufacturing for repetitive, boring, tedious factory chores (Freund, 1982). Additionally, the effects of modern microelectronics will be to lower cost, improve performance, and substantially widen the availability of automation technology.

FIGURE 10-1. Man and the machine. Charlie Chaplin in *Modern Times. Courtesy of The Museum of Modern Art/Film Stills Archive, New York City.*

Automation and U.S. Employment

Productivity improvements resulting from the use of robotics and automated manufacturing techniques can affect labor in a number of ways, depending on the following factors:

1. The relative proportion of machinery to workers (the capital-to-labor ratio) in a given industry.

2. The extent of change in prices and production volumes for U.S. firms.

3. The supply of qualified workers with specific job skills in a given industry.

In the past, despite some initial displacement of labor, labor-saving techniques have ultimately led to improved living standards and higher real wages. Also, since total employment is a function of real economic growth and robots can have a positive effect on real economic growth, they could, therefore, have a positive effect on total employment. U.S. employment in a given industry may rise if productivity improvements are combined with increases in production volume. Effective labor compensation may also rise if productivity improvements lead to shorter work weeks and better profit margins as a result of improved production volume and profitability. In addition, average wage levels will change with changes in the necessary mix of workers' skills resulting from the implementation of robotics; with proper retraining, workers will have the opportunity to assume better positions within the industry and thus gain a higher wage level. The working environment can also be improved by using robots for processes that create hazardous working conditions.

One great opportunity afforded by robots for increased employment in the United States is the return of electronic and semiconductor assembly work to the United States. This can be achieved by overcoming the advantages of low wage rates overseas through a combination of automated manufacturing equipment and domestic labor which will yield a total cost per unit competitive with overseas costs. There is potential for bringing back several hundred thousand jobs by enabling U.S. industry to compete with existing low-wage labor in Mexico and the Far East. This approach is particularly applicable in the assembly of semiconductor devices and electronic assemblies. Many suggest that U.S. assembly plants could not only reduce the distance between semiconductor wafer fabrication and packaging operations, but also reduce turnaround times, cost inventories, and investment risks. The need for domestic assembly work also promises to increase as the electronics industries place more emphasis on low-volume semicustom and custom components.

Other industries in which the United States could find itself capable of competing in fields previously lost to low-wage workers overseas are the development and assembly of black and white television sets and consumer electronics devices such as radios and recorders.

How much benefit will be derived from the use of robotics and how we choose to distribute this benefit is a policy question which should be addressed in the early stages of the development of the technology. Will employers keep the profits of increased productivity or will they expand production? Worker cooperation will depend upon how early robotics programs answer that question. Worker resistance could slow the pace at which American plants automate, and therefore jeopardize America's ability to compete in world markets. The introduction of robots may not displace workers if U.S. companies emulate the Japanese practice of making long-term investments in workers and capital equipment alike. Had automotive workers been permitted greater input into production decisions as they are in Japan, many believe that Detroit might be in better shape today.

It is recognized by many that it is the relationship of new technology and worker motivation that leads to productivity growth. Technology cannot serve as a substitute for good management and human motivation. In many cases today, the skill of the worker is undervalued. Worker involvement in the design of the manufacturing process and the work place itself would offer new opportunities for the enhancement of motivation.

Worker and Union Fears and Attitudes

In many forums, the debate over robots has pitted the robot against the worker in a winner-take-all fight with the greater efficiency of the robot attempting to offset the human creativity of the worker. The advantage of the robot is seen as a disadvantage to the worker. Workers and unions fear the loss of jobs and the devaluation of worker skills, or having to accept lower wages or increased output to prevent the wholesale loss of jobs to an army of robots. In general, the wish is for participation in the job design process and guaranteed job security. Unions desire contracts based on natural attrition clauses that state that the introduction of technology will be keyed to the rate at which people retire or move to other jobs.

Even Japan's robot invasion has not gone entirely unopposed. Under pressure from the unions, Prime Minister Zenko Suzuki has ordered the Labor Ministry to study the robots' impact on the labor market and working conditions. The results will not be in for nearly two years, but according to a recent poll by Nikkei Business magazine, 97 percent of in-house unions and 79 percent of management think robotization will lead to increased unemployment. A recent survey by the Federation of Metalworkers Unions revealed that, while 52 percent of its member unions welcome robots on production lines, fully 45 percent said they didn't know what to think—a neutral vote widely interpreted as an indication of increasing anxiety. Says Kazuyoski Tsuda, an official at the federation: "I can feel a silent uneasiness growing" ("Here Come the Robots," 1982).

Worker resistance in Japan has stymied robot introduction in cases where workers felt that robots were taking away the pleasant and easy jobs, leaving dirty and unpleasant work for people. However, where robots have replaced people in hostile working environments, such as welding, painting, cutting, grinding, and heavy materials handling, Japanese companies have found that workers learn to live with robots and even come to appreciate their value. Furthermore, if workers are retained as supervisors of robots, they often take great pride in the robots now working for them. According to a survey by Kikkei Mechanical, about two-thirds of the Japanese companies which have installed robots use line workers to teach and maintain them (Ohmae, 1982).

When management pays careful attention to employees' career paths, job enrichment, and job assurance, and offers comprehensive retraining programs, blue-collar workers learn to live with robots and eventually a peaceful man-machine interface is established.

Attitudes of Management

One of the influential deterrents to more rapid adoption of robots has been managerial concern about the reaction of the labor force. The introduction of robots to replace operators in dangerous or especially uncomfortable environments was readily accepted as was their use in unduly strenuous jobs. The use of robots in highly routine jobs has also been commonly accepted by labor, provided that the replaced operators were given other assignments. But there seems to be widespread concern among managers that robot installations which threaten substantial employment reductions in existing plants may well engender serious labor problems which would be likely to reduce expected cost-savings substantially.

Accordingly, major installations are likely to be restricted to new plants which can establish manning levels in accordance with their new operating characteristics. Until the aforementioned deterrent is overcome, the diffusion of robot technology to older plants will be delayed. Only when an immediate threat to the survival of the plant is recognized by labor is resistance unlikely to inhibit the introduction of new technologies.

Major incentives can be utilized to achieve labor's cooperation in the use of automated manufacturing systems. For example, the sharing of increased productivity benefits with the workers is one useful approach; another is to contractually allow for retraining programs in the case of labor displacement by automation.

Another big problem is fear and the avoidance of risk—the risks of making a decision and upsetting production, with the consequent risk to one's job. One often hears, "It won't work in my plant!," a reaction due solely to the fear of trying the equipment because of anticipated disruptions and slow-downs in the production process. Unfortunately, the U.S. business system is often dedicated to the very short term.

Labor Displacement

There is historic evidence that automation has led to overall increased employment in the past; nevertheless, the introduction of automation initially led to significant labor displacement. By 1812, after the introduction of Hargreave's "jenny," one spinner in a mill could produce as much in a given period as 200 workers could have produced at home using a spinning wheel. Employment in the British factories of cotton textile industry increased from less than 100,000 in 1770 (Hargreave's patent date) to about 350,000 in 1800 (Deane, 1979). However, at the same time the cottage industry of the home spinners dropped significantly.

Another example of significant change and labor displacement is the transition due to the increased utilization of technology on the U.S. farms. Agricultural workers declined from one-half of the nation's employed in 1900 to less than five percent today (Kalmbach, 1982). The number of farm workers in the U.S. decreased from 10 million in 1950 to 3 million in 1975. Over the same period, agricultural productivity increased by a factor of 7 (Dorf, 1974). During this period the worker left the farm to migrate to the city and seek new opportunities in industry.

A number of industries, such as the automobile and farm machinery industries, are currently beset with reduced labor requirements which may continue even after the 1982 recession is overcome. Throughout the period 1950 to 1982, employment in manufacturing declined from 33.7% to 21.4% as a percentage of all nonfarm jobs (Sugarman, 1980).

Some alarming predictions of labor displacement due to robots are beginning to appear. One industrial official is quoted as stating that it would be possible to replace one-half of his company's 37,000 assembly workers with machines ("Where the Jobs Are . . . ," 1981). It may be possible by 1990 to displace up to 10% of those employed in industry as painters, welders, and assemblers. One pessimistic estimate is that 100,000 jobs may be lost in the U.S. automobile industry over the next decade (Shaiken, 1980). General Motors reports that one robot displaces 1.7 workers in an assembly plant and 2.7 workers in a manufacturing plant (Main, 1982).

A recent study by the Congressional Budget Office predicted that the number of workers displaced by automation and the economic problems of numerous declining industries would reach 2 million by the end of 1982 (Ayres, 1981). They state: "Although the degree of actual dislocation will be somewhat less than the number of jobs lost due to natural attrition and industrial retraining efforts, dislocation is likely to be substantially larger than in the previous period of rapid technological change begun in the 1950's."

A study by the Robot Institute of America of the Society of Manufacturing Engineers estimates that 440,000 workers will be replaced by robots by the end of this century. However, the study also estimates that all but 20,000 can expect to be accommodated by attrition or by allowing for retraining of the operators for new positions ("Organized Labor . . . ," 1982).

Another study accomplished at Carnegie-Mellon University (Ayers, 1981) forecasts significant labor displacements. While attrition will account from many displaced workers, it is expected that hundreds of thousands of workers will be required to transfer to new, growing sectors of industry such as the information industries. The current percent of the U.S. work force in information-related employment is shown in Figure 10-2. Note that even Venezuela has a significant percent of its workers in information-related industries.

However, transfer of workers to new industries usually involves relocation and retraining, and the effects on an individual who has been replaced by automation can be traumatic.

As capital equipment is substituted for labor in heavy industries such as auto and steel, one can expect reduced employment in those manufacturing enterprises where the robot directly replaces the human worker. For example, a painting robot may replace a human painter on an automobile production line. Similarly, robots used to assemble and

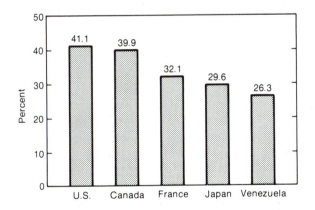

FIGURE 10-2. With 41.4 percent of its work force engaged in information-related employment, the U.S. is the leader of the five nations represented in the graph. *Source: U.S. Congressional Committee on Home Administration.*

insert electronic devices on a printed circuit board replace individuals. An example is Pioneer Electronic Corporation's action to cut the cost of installing electronic parts in circuit boards by 75% by turning the work over to robots. Pioneer now uses about 100 robots (Cook, 1981).

Labor displacement will be a critical issue if robotics and automated manufacturing are rapidly applied in the U.S. However, it has been recognized in Detroit and elsewhere that if firms are unable to maintain the productivity and quality of products, and therefore retain an important share of the international market, they will atrophy and decline, thus leading to even greater labor displacement and unemployment. In order for a firm to remain competitive it must invest capital in equipment, thus increasing productivity per worker and therefore be able to compensate workers based on a fair return. The only way that employment in an industry can be maintained constant is for that industry to be emerging and growing. The United States is fortunate that it has industries which have the potential to innovate and economically develop and grow, such as electronic assembly, agricultural food processing, semiconductor assembly and manufacturing, and aircraft manufacturing. One could even argue that the auto and steel industries of the nation could experience increased labor utilization if market share could be retained or improved in competition with international manufacturers.

Although labor displacement in certain industries may be inevitable, it should be remembered that the estimated number of jobs that could be performed by robots by 1990 may be less than 10 percent of all jobs, and could be less than 5 percent. In addition, it is hoped that most workers displaced by robots would be spared unemployment because of retraining and retirement.

New Challenges in the Work Place

The use of robots in the industrial work place offers new opportunities for the partnership of labor and management. The current primary use of robots is for boring, repetitive tasks. One precondition for these uses is that the operations are short-cycle, repetitive ones, which means that employees affected by the use of robots are in general those from whom only minimal skill levels are required. Complex operations which are not subject to schematized repetition are usually not feasible for the use of robots. The use of robots changes the quantity and quality of more highly technical operations, but this is not due to direct substitutions.

Whether the work situation of employees improves or deteriorates through the employment of robotics depends substantially on which manual work functions are mechanized, which functions remain, and how these are distributed. It is often asserted that the application of robots results in improvement of working conditions, reduction of hard physical work, and liberation of workers from monotonous and environmentally stressful jobs. Although the introduction of robots and automation may mean a decrease in certain kinds of work stress, the level of technique and the organization of work can cause new stresses to arise. Processes of technical rationalization do not necessarily equate to a reduction of stress in production work or with an improvement in working conditions.

The rhythm of work at a work place has a strong impact on the workers, and contrary to the expectation that the introduction of industrial robots would raise the worker's mobility margin, in many cases the use of robots in the manufacturing process has caused the rhythm obligation for periphery workers to become more rigid (Kalmbach, 1982). Nevertheless, it is feasible to design the technical and work place organization so that the worker can be freed from the work rhythm of the system.

Safety measures in the work place are required to protect the worker. Two recent fatalities in an automated work place involved robots, the first of which happened at a Kawasaki plant when a 37-year-old engineer, with more than 10 years' experience with robots, was crushed to death. He had found a malfunctioning process machine, stepped over a safety rope to inspect and correct the problem and was struck from behind by the robot. The second accident happened at a Japanese automobile plant when a visiting engineer inspecting a sequence robot was killed by it ("Industrial Robots Can Be Dangerous," 1982).

Another consideration is that the use of robots may intensify the tendency towards social isolation of the worker. The monotonous, isolated tasks of machine-minders may lead to boredom and loss of a sense of participation.

Generally, however, when human labor is freed from the boring, repetitive, physically exhausting jobs, the trend is more often towards a more desirable work environment. Work places characterized essentially by heavy physical and environmental stresses can be improved with the addition of robots, but one must avoid the development of new, residual working places in which the employees are confronted with a higher rhythm obligation, greater intensity of output, and a heightened social isolation.

Education and Training Requirements

Workers traditionally have been left to their own resources when they find new technology has affected or eliminated their jobs. However, if the extensive introduction of robots and automated manufacturing is to succeed, retraining programs for displaced workers need to be implemented ("Retraining Displaced Workers . . . ," 1982). Retraining displaced blue-collar workers is important to successful incorporation of robots. Many companies in Japan and the U.S. are converting displaced workers to maintenance personnel. It has been predicted that by 1990 half of the workers in U.S. factories will be white collar specialists able to maintain and service robots ("Robots to Oust Many Workers . . . ," 1982).

The nation and industry must provide for the retraining of workers displaced by automation using college programs as well as industrially sponsored programs. In addition, an appropriate agency could maintain information on displacement and labor needs as a result of automated manufacturing throughout the nation, and assist in the transfer of displaced labor to other firms.

The labor needs of the robot and computer-aided manufacturing industry cut across all areas of labor. The industry requires designers, engineers, maintenance technicians, and on-floor supervisors and operators of the equipment. All of these individuals must be capable of understanding at various levels the operation of the robot and the electrical, hydraulic, and pneumatic devices which are utilized in such equipment.

There is an especially great need for manufacturing engineers, a field which was of interest and importance in the universities between 1930 and 1950 but which was allowed to atrophy; today American universities offer few if any programs in manufacturing. Since robotics is a new and emerging field, very few graduates of engineering colleges are knowledgeable in the fields of control systems, computers, and hydraulic and pneumatic equipment. They therefore find it difficult to effectively utilize the available equipment and are unable to handle the interdisciplinary requirements of robotics. On the positive side, there are a great number of engineers graduating in the field of computer science and engineering who will be able to develop new software packages to support automated manufacturing equipment.

There is also a great need for properly educated technicians who are able to maintain and assist in the operation and installation of robotics and automated manufacturing equipment. There are only a limited number of programs in the community and technical college system that bridge the fields of electronics, pneumatics, and hydraulic systems and also include some discussion of the computer software included in the systems.

Over the next 5 years there may be a need in the U.S. for several thousands of engineers as well as thousands of support technicians. Given the fact that the universities and community colleges are currently unprepared or unable to meet this need, a concerted drive will be required to support the robot and automated manufacturing industry if it is to develop its leadership capabilities.

11

The Future of Robotics and Automated Manufacturing

By the year 1990, there may be over 100,000 robots installed in the United States. The form, design and technology of these robots will improve as time progresses.

Before robots can significantly impact productivity of the economy as a whole, they must be used in hundreds of thousands of applications. This will be made possible as a number of technical problems are overcome.

The first technical need is the improvement of the accuracy of positioning the robot. Although the repeatability of most robots is in the order of .05 inch over its working volume, and in some cases as good as .005 inch, the absolute positioning accuracy may be off as much as .250 inch in some regions of the working envelope. Thus, it is not currently possible to program a robot to go to an arbitrary mathematically defined point in a coordinate space and have any assurance that the robot will come closer than a quarter of an inch. This creates major problems in programming a robot from a computer terminal or in transferring programs from one robot to another. Through better calibration procedures and improved mechanical systems, the arbitrary accuracy of the system will eventually be improved, but until then, robots for use in the assembly of small batches of products will be relatively uneconomical (Smith & Wilson, 1982).

Another area needing improvement is the dynamic performance of robots. In many assembly tasks, present-day robots are too slow and clumsy to effectively work in cooperation with human labor. In addition, we are not achieving any magnitude of improvement in assembly tasks using robots in contrast to human performance. Robot structures are typically quite massive and unwieldy. Most robots can only lift about one-tenth of their own weight and many cannot do that. New mechanical designs using lightweight materials and advanced structures will be necessary to achieve high speeds.

Much also remains to be accomplished in the design of end effectors and grippers. Typically, robot hands today consist of only pinched grippers with one degree of freedom, that is, open or shut; by contrast, the human hand has five fingers, each with four degrees of freedom and an opposing thumb. No robot has come close to duplicating the dexterity of the human hand.

Sensors of many different kinds also need to be developed. Robots must become able to see, feel, and sense the position of objects in a number of different ways. Robot sensors will be developed using optical, x-ray, and acoustic detectors, among others. Other robot sensory input such as touch and force appear to be necessary. In addition, control systems are needed which can process sophisticated sensory data from a large number of different types of sensors simultaneously. Robot control systems need to have much more sophisticated internal models of the environment in which they work. Future robot control systems will have data bases similar to those of computer-aided design (CAD) systems.

Techniques for developing software which will control the robots must be vastly enhanced. Programming by teaching the robot is impractical for small-lot production, especially for complex tasks where sensory interaction is involved. Eventually, it will be necessary to have a whole range of programming languages and debugging tools at each level of the sensory control hierarchy. Trends in the field of computer-aided manufacturing (CAM) are toward distributed computing systems where a large number of control computers, robots, sensors, and machine tools all interact and cooperate as an integrated system. This represents an extensive challenge in the area of software design and computer interface ("Robots Make Slow Advance . . . ," 1981).

In addition, many potential robot applications will require robot mobility. Robots will be used along the sides of ships for welding and cleaning; there will be need to work from scaffolding and on the sides of buildings, as well as to move throughout factories.

There is also a need for research on high-speed manipulation, particularly for industrial applications. Some of the difficulties in controlling high-speed manipulators are that the inertias can change with the arm configuration and with the payload, such as on the space shuttle. Structural flexibility complicates dynamic control, especially if it interacts with the discrete-time effects of a digital controller. New manipulator materials which are rigid and lightweight, such as composites, merit investigation. Using flexibility to attain high-speed motion (for example, the movement of a fly fisherman's rod) is another possible approach, but such manipulators would require control algorithms which are far more advanced than any in use today.

Typically, manipulators have six axes to completely position and orient objects. However, research is needed on redundant axis manipulators to aid in working around obstacles. Research is also needed on designing fault-tolerant robotic systems which have the capability of graceful degradation; that is, systems which continue to operate in a degraded mode.

A central problem with industrial robotic intelligence is that a robot may have a large number of alternative actions, but the user does not want to be burdened with specifying them. This problem can be addressed by research work on interfacing robots with existing information systems, on designing products for automation, on the off-line simulation of robots with sensory feedback, and on restart procedures.

The problems of manipulator design, sensing, control, and planning should be set into common framework. The representation of the world, objects, and actions are central issues, along with planning and executing complex actions. The question of how to coordinate multiple activities and how to specify robot actions is a matter for future research.

Technical advances will be required with regard to end effectors, robot speed, cost, and design. A recent technology forecast carried out by the Society of Manufacturing Engineers predicts many significant developments as summarized in Table 11-1. This forecast was achieved utilizing a sixty-member Delphi panel to obtain a composite of expert opinion (Smith & Wilson, 1982). The robot available by 1990 is envi-

TABLE 11-1.   A Summary of the SME Delphi Forecast

Forecasted Robot Characteristics by 1990	Percent of Robots Sold in 1990 Having the Forecasted Characteristics
Robots made of modular, standard components	63%
Use of a general-purpose hand	20%
Self-propelled robots	15%
Repeatability less than or equal to .001 inch	93%
Use of integral sensors	60%
Use of pattern recognition and feedback	22%
Use of vision capability	25%
Robots that will be programmed on-line	20%

sioned as modular, accurate, intelligent, and incorporating sensors for control.

In the coming decade we can look for systems with improved repeatability and reliability, lighter weight, lower cost and more extended sensor capabilities. In addition, systems will more greatly rely on the combination of sensors, hierarchical control, and the computer.

Future Manufacturing Systems

The interactive system of robots, computers, and machines in a manufacturing system will develop over the next decade. One schematic of the factory of the future is shown in Figure 11-1. The interaction of computer-aided engineering, computer-aided manufacturing, and intelligent warehouse systems will make up the factory of the future. Both the computer-aided manufacturing system and the warehouse system incorporate fixed robots, CNC, and mobile robots. An example of an unmanned machining center is shown in Figure 11-2.

One version of the manufacturing system of the future would include the design of programmable manufacturing cells composed of flexible machines, manipulators, and vision systems. Also, the CAM system may include knowledge-based systems, which is the application of artificial intelligence principles to management systems in order to make them as flexible and responsive as machine processes.

Future Applications

Undersea Explorations.   One of the major potential uses of robotic devices is performing tasks in environments in which humans cannot operate safely or efficiently. One such environment is under water, where robots can be used for construction and repair of structures, retrieving lost objects, gathering geological samples, exploring the depths for minerals, and operating underwater mining and drilling equipment. Undersea functioning will require on-board computer processing for sensing and mobility far greater than is currently available. Underwater robots could be virtually weightless, with their buoyancy controlled by filling or emptying air chambers. They could be maneuvered by propellers or by jets of water; even walking underwater could easily be accomplished by a two-legged robot.

Space Applications.   The distance we can travel in space exceeds the distance over which we can exert real-time controls. Therefore, the

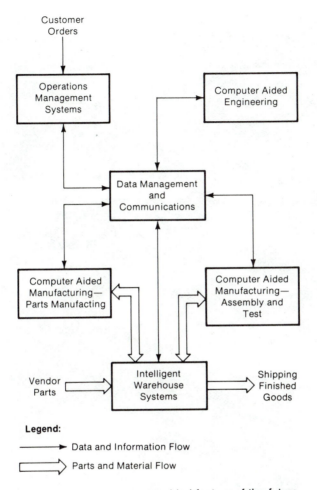

FIGURE 11-1. A computer-aided factory of the future.

devices we send into space must either contain their own operating intelligence which can be preprogrammed, as with the Viking explorers, or else be flexible and adaptive, as with robots. Robotic devices are very useful in carrying out such tasks as construction in space. Robotic machines with artificial intelligence will be capable of repairing themselves and building other machines. Some engineers have proposed a plan for using robots on the moon for exploration and mining, including a factory for self-replication of robot modules (Garrett, 1982; von-Tiesenhausen & Darboro, 1981).

FIGURE 11-2.   Kearney and Trecker unmanned machining center.

Mining.   A mine shaft is the nation's most hazardous work site, yet the variability in mining conditions has resisted automation, requiring human operators at the most hazardous points in the extraction process. By incorporating sensory faculties and logic into mining equipment, many of the hazards and adverse working conditions of mining can be eliminated by the use of robots as substitutes for human miners. Research on such systems is devoted to developing direct remote control based on extensive use of sensory systems, in which a degree of decisionmaking power is built into each machine. The ultimate goal is a completely autonomous robot mining system, with machine intelligence carrying out sensing, planning, and control operations.

Medical Applications.   Technologies evolving from robotic research will find several kinds of medical applications. The most obvious is in prosthetics. Flexible hands with touch sensitivity are being developed and new principles will find ready application in improving artificial limbs. Research in applying artificial intelligence and robotic devices to extend the ability of physically handicapped people to control their environment is underway. Advances in multisensor systems and the related problem of interpreting a situation based on multisensory inputs can contribute to advances in microsurgery.

Construction.   It will probably be many decades before robots are able to do all the tasks necessary to build a house. Nevertheless, by 1990 use of robot labor in certain construction processes may significantly affect the cost of new housing. Once construction robots become dextrous, inexpensive, and simple to use in a large variety of practical construction tasks, labor-intensive construction techniques that have been abandoned because of prohibitive costs may be once again economically practical.

The amount of software needed to enable a robot to perform routine construction and assembly tasks in a relatively unstructured environment is large. However, once the software is properly structured it can be easily transferred from one robot to another and from one type of robot to another. Such software is task-specific; each task requires a unique program—indeed, an entire set of programs—at each level in the sensory-control hierarchy. Nevertheless, several different tasks utilize the same set of subtasks, merely arranged in differing orders.

Household Robots.   The 1990's could be the decade during which the household robot becomes practical. Once plastic robots are fully developed and become inexpensive in industrial applications, and once advanced software and sensory systems are developed for construction robots, the same technology can be applied to the development of household robots. The environment of the home is as variable and complex as that of the construction site. Each home has a different floor plan and arrangement of furniture. In order for a household robot to negotiate through the average room, it must have an internal map of the furniture placement and permissible pathways. A household robot must know where the dishwasher is, where the cabinets are located, and where each of the various types of plates and utensils is to be stored. If it is to vacuum, it must have a map of what to vacuum as well as a sensory system that can recognize the difference between patches of dirt and patterns in a rug (Albus, 1981). And yet the cost of a household robot would probably need to be less than the cost of an automobile in order for it to be marketable.

International Competition

The Japanese predominance lies largely in that country's leadership in robot application. There are more than 70 Japanese robot manufacturers, about twice as many as in West Germany, and three times more than in the U.S. In addition, the Japanese have the two plants in the world that come closest to the ideal of totally automated and integrated manufacturing, with computerized design, manufacturing, and control replacing human functions. Given the complex and shifting requirements of total manufacturing integration, purchasers of robots will put a premium on integrated solutions to their factory automation problems, and so will favor companies that provide that service.

Robot users also want standardization, but American industry is also lagging here. Unlike Japan, the U.S. has no national plan for fostering robotics, and therefore no government-inspired standardization. And, unlike IBM in the computer industry, there is no one company powerful enough to establish norms. Unless U.S. firms pool their efforts and emerge with vertically integrated companies, it may be the Japanese robotics firms that will dictate industry standards. The Japanese appear to favor the application of proven and easily attainable solutions, and produce products which are almost always reliable. They are also quick to get their goods into use, hence gaining application experience, while at the same time establishing a market foothold.

It is reasonable to infer that the lack of success of the U.S. and European robot industry is due to factors other than competitors' superiority. With renewed economic growth and capital investment, the competitive abilities of Western robot and automation firms could well match those of the Japanese firms.

Social and Economic Consequences

Most observers agree that the largest social consequence of the rapid introduction of robots may be a large displacement of industrial workers, who would require retraining and reemployment in order to share in the increased wealth of the nation.

Even in an economically efficient society, demand can always be increased. All that is necessary is for increased spending power to flow to consumers at the same rate that increased output flows from the factories. This, however, is not always easy to achieve in a politically acceptable way when wages are tied to the amount of work input and not to the level of output. This structural flaw in the income distribution system creates at least the appearance of a threat of unemployment which is serious enough to delay, if not prevent, the rapid development of a massive robot labor force from becoming an acceptable goal of national economic policy. As long as the unemployment rate for human workers is unacceptably high it will be extremely difficult to generate significant public enthusiasm for massive investments in robots.

Robot technology offers the possibility of significantly reducing poverty, not only in the United States, but all over the world. The cure for poverty is wealth, and robots have a virtually unlimited capacity to create wealth. Robots are almost like a new race of creatures, willing to be our slaves, who will work merely for the cost of their purchase price and their upkeep. The self-reproductive potential of robots suggests that they might be manufactured in large quantities at exponentially declining costs.

As we consider the possibility of self-replicating robots we recognize that we may be endowing machines with the one attribute that we thought was reserved to living organisms—self-reproduction. It's true that it will take not only robots but other assorted electronic automatic machines to make a robot, but the fact remains that machines may eventually be capable of producing new machines without human intervention. If such self-replicating machines were also endowed with artificial intelligence of an advanced nature, society may be faced with a new challenge: What do humans do to guard against eventual domination by intelligent machines? This has been an issue for mankind for many decades and may come to be of central importance in the decades ahead.

Glossary

Accuracy   The ability of the manipulator to position the end effector (tool or gripper) at a specified point in space upon receiving a command by the controller.

Actuator   A transducer that converts electrical, hydraulic, or pneumatic energy to cause motion of the robot.

Arm   An interconnected series of mechanical links and joints that support and move the end effector through space.

Artificial Intelligence   The ability of a machine to perform certain complex functions normally associated with human intelligence, such as judgment, pattern recognition, understanding, learning, planning, and problem solving.

Base   The platform which supports the manipulator arm.

Closed Loop Control   Robot control which uses a feedback loop to measure then compare actual system performance with desired performance, and then makes adjustments accordingly.

Computer-Aided Design (CAD)   The use of a computer to assist in the design of a product or manufacturing system.

Computer-Aided Manufacturing (CAM)   The use of a computer to assist in the manufacturing process.

Computer Numerical Control (CNC)   The use of a dedicated computer within a numerical control unit that provides data input for the machine.

Contact Sensor   A device that detects the presence of an object or measures the amount of force or torque applied by the object through physical contact with it.

Controller   The robot brain, which directs the motion of the end effector so that it is both positioned and oriented correctly in space over time.

Degrees of Freedom   The number of independent ways in which the end effector can move, defined by the number of rotational or translational axes through which motion can be achieved.

Direct Numerical Control (DNC)   The use of a computer for providing data inputs to several remote numerically controlled machine tools.

End Effector   The tool or gripper which is attached to the mounting surface of the manipulator wrist in order to perform the robot's task.

External Sensor   A feedback device for detecting locations, orientations, forces, or shapes of objects outside of the robot's immediate environment.

Flexibility   The ability of a robot to perform a variety of different tasks.

Force Sensor   A device that detects and measures the magnitude of the force exerted by an object upon contacting it.

Gripper   The hand of the manipulator, which is used by the robot to grasp objects.

Group Technology   The grouping of parts into categories having common characteristics, such as shape, so that all parts within each category can be processed together.

Hard Automation   Automated machinery that is fixed, or dedicated, to one particular manufacturing task throughout its life.

Interface   A boundary between the robot and machines, transfer lines, or parts outside of its immediate environment. The robot must communicate with these items through input/output signals provided by sensors.

Internal Sensor   A feedback device in the manipulator arm which provides data to the controller on the position of the arm.

Jointed Arm Robot   A robot whose arm consists of two links connected by "elbow" and "shoulder" joints to provide three rotational motions. This robot most closely resembles the human arm.

Leadthrough Programming   A means of teaching a robot by leading it through the operating sequence with a control console or a hand-held control box.

Load Capacity   The maximum weight that can be handled by the robot without failure.

Manipulator   The mechanical arm mechanism, consisting of a series of links and joints, which accomplishes the motion of an object through space.

Manual Programming   A means of teaching a robot by physically presetting the cams on a rotating stepping drum, setting limit switches on the axes, arranging wires, or fitting air tubes.

Microprocessor   A compact element of a computer central processing unit, constructed as a single integrated unit and increasingly used as a control unit for robots.

Non-Servo Control   The control of a robot through the use of mechanical stops which permit motion between two end points.

Numerical Control (NC)   A means of providing prerecorded information that gives complete instructions for the operation of a machine.

Off-Line Programming   A means of programming a robot by developing a set of instructions on an independent computer and then using the software to control the robot at a later date.

On-Line Programming   A means of programming a robot on a computer that directly controls the robot. The programming is performed in real time.

Payload   The maximum weight that can be handled by a robot during normal operation.

Pick-and-Place Robot   A non-servo robot which operates by moving along each axis between two end points. These robots are generally used for simple part transfer operations.

Point-to-Point Motion   A type of robot motion in which a limited number of points along a path of motion is specified by the controller, and the robot moves from point to point rather than in a continuous, smooth path.

Programmable   A feature of a robot that allows it to be instructed to perform a sequence of steps, and then to perform this sequence in a repetitive manner. It can then be reprogrammed to perform a different sequence of steps, if desired.

Reach   The maximum distance from the center line of the robot to the end of its tool mounting plate.

Reliability   The percentage of time during which the robot can be expected to be in normal operation (i.e., not out of service for repairs or maintenance). This is also known as the "up time" of the robot.

Repeatability   The ability of the manipulator arm to position the end effector at a particular location within a specified distance from its position during the previous cycle.

Robot   A reprogrammable multifunctional manipulator designed to move material, parts, tools, or specialized devices through variable programmed motion for performance of a variety of tasks.

Rotational Motion   A degree of freedom that defines motion of rotation about an axis.

Sensor   A feedback device which can detect certain characteristics of objects through some form of interaction with them.

Shoulder   The manipulator arm link joint that is attached to the base.

Speed   The maximum speed at which the end of the manipulator arm can move at a certain load.

Swing   The rotation about the center line of the robot.

Tactile Sensor   A sensor that detects the presence of an object or measures force or torque through contact with the object.

Teaching   The process of programming a robot to perform a desired sequence of tasks.

Touch Sensor   A sensor that detects the presence of an object by coming into contact with it.

Vertical Stroke   The amount of vertical motion of the robot arm from one elevation to the other.

Vision Sensor   A sensor that identifies the shape, location, orientation, or dimensions of an object through visual feedback, such as a television camera.

Walkthrough Programming   A method of programming a robot by physically moving the manipulator arm through a complete operating cycle. This is typically used for continuous path robots.

Work Envelope   The three dimensional space that defines the entire range of points which can be reached by the end effector.

Wrist   The manipulator arm joint to which an end effector is attached.

Bibliography

REFERENCES

Agin, G. J., and Duda, R. O. *SRI vision research*. Proceedings of 2nd USA-Japan Computer Conference, Tokyo, 1975, 113–117.

Albus, J. S., *Brains, behavior and robotics*. New York: McGraw-Hill, 1981.

"Artificial intelligence." *Business Week*, March 8, 1982, 66–75.

"Automatic factory: Identify its fingerprints." *Material Handling Engineering*, June 1981, 92–96.

Ayres, R. V. *The impact of robotics on the workforce and workplace*. Pittsburgh: Carnegie-Mellon University, June 1981.

Ball, R. Europe's durable unemployment woes. *Fortune*, January 11, 1982, 66–74.

Ballard, D. H., and Brown, C. M. *Computer vision*. New York: Prentice-Hall, 1982.

Beercheck, R. C. Fluid power. *Machine Design*, January 21, 1982, 58–64.

Bulkeley, W. M. First intelligent mobile robots soon may serve as plant sentries. *Wall Street Journal*, August 16, 1982, 15.

Burck, C. G. Can Detroit catch up? *Fortune*, February 8, 1982, 34–39.

Cook, D. T. Why Japan leads U.S. in robot race. *The Christian Science Monitor*, December 8, 1981, 11.

Cousineau, D. T. Robots are easy, it's everything else that's hard. *Robotics Today*, Spring 1981, 28–35.

Dallas, D. B. The advent of the automatic factory. *Manufacturing Engineering*, November 1980, 66–75.

Davis, R. Expert systems. *The AI Magazine*, Spring 1982, 3–22.

Deane, P. *The first industrial revolution*. New York: Cambridge University Press, 1979, 90–91.

Dorf, R. C. *Technology and man*. San Francisco: Boyd and Fraser, 1974.

Dorf, R. C. *Introduction to computers and computer science*, 3rd Edition. San Francisco: Boyd and Fraser, 1981.

Dorf, R. C. *Modern control systems*, 3rd Edition. Reading MA: Addison-Wesley, 1981.

Engelberger, J. F. *Robotics in practice*. New York: AMACOM, 1980.

Flint, J. The myth of the lazy American. *Forbes*, July 6, 1981, 105–111.

Freund, W. C. The looming impact of population changes. *Wall Street Journal*, April 6, 1982, 28.

Froehlich, L. Robots to the rescue. *Datamation*, January 1981, 85–96.

Garrett, R. C. The inventor's sketchpad. *Interface Age*, February 1982, 27–33.

Glorioso, R. M., and Osorco, F. C. *Engineering intelligent systems: Concept, theory, and applications.* Billerica ME: Digital Press, 1980.

Haitt, B. Toward machines that see. *Mosaic*, December 1981, 2–8.

Harmon, L. D. *Automatic tactile sensing.* Proceedings of the Robot IV Conference, Society of Manufacturing Engineers, March 1982.

Hartley, J. New robot designs give increased production. *The Industrial Robot*, September 1980, 186–188.

"Here come the robots." *Newsweek*, August 9, 1982, 58.

Hill, J. W. MiniMover 5. *Robotics Age*, Summer 1980, 18–27.

Hudson, C. A. Computers in manufacturing. *Science*, February 12, 1982, 818–825.

Huston, R. L., and Kelly, F. A. The development of equations of motion of single-arm robots. *IEEE Transactions on Systems, Man and Cybernetics*, June 1982, 259–265.

"Industrial robots can be dangerous." *American Industry of Plant Engineers Journal*, Summer 1982, 28.

Iversen, W. R. Vision systems gain smarts. *Electronics*, April 7, 1982, 89–90.

Jarvis, R. A. A computer vision and robotics laboratory. *Computer*, June 1982, 8–23.

Jurgen, R. K. Detroit bets on electronics to stymie Japan. *IEEE Spectrum*, July 1981, 29–32.

Kalmbach, P. Robots effect on production, work and employment. *The Industrial Robot*, March 1982, 42–46.

Krouse, J. K. Smart robots for CAD/CAM. *Machine Design*, June 25, 1981, 85–91.

Krouse, J. K. CAD/CAM broadens its appeal. *Machine Design*, April 22, 1982, 54–59.

Kruger, R. P., and Thompson, W. B. *A technical and economic assessment of computer vision for industrial inspection and robotic assembly.* Proceedings of the IEEE, December 1981, 1524–1538.

Lawrence, B. Ford plant's robots meet the press. *Sacramento Bee*, May 13, 1982, Cl.

Lerner, E. J. Computer-aided manufacturing. *IEEE Spectrum*, November 1981, 34–39.

Link, C. H. MBB-Europe's automated factory showcase. *Assembly Engineering*, June 1981, 53–54.

Main, J. Work won't be the same again. *Fortune*, June 28, 1982, 58–65.

Mangold, V. The industrial robot as transfer device. *Robotics Age,* August 1981, 20–26.

Marsh P. America's factories race to automation. *New Scientist,* June 25, 1981, 845–847.

McCormick, D. Making points with robot assembly. *Design Engineering,* August 1982, 24–28.

Meacham, J. TIG welding with robots. *Robotics Age,* March 1981, 28–31.

Motiwalla, S. Continuous path control with stepping motors. *Robotics Age,* August 1981, 28–36.

Nevatia, R. *Machine perception.* New York: Prentice-Hall, 1982.

Nevins, J. L., and Whitney, D. E. Computer-controlled assembly. *Scientific American,* February 1978, 62–74.

Ohmae, K. Steel collar workers: The lessons from Japan. *Wall Street Journal,* February 16, 1982, 25.

"Organized labor faces the robot age." *Best of Business,* April 1982, 49.

Ottinger, L. V. A plant search for possible robot applications. *Industrial Engineer,* December 1981, 26.

Ottinger, L. V. Evaluating potential robot applications in a system context. *Industrial Engineer,* January 1982, 80–87

Owen, A. E. Economic criterion for robot justification. *Industrial Robot,* September 1980, 176–177.

Paul, R. P. *Robot manipulators.* Cambridge MA: MIT Press, 1981.

"Pentel signs its name to the high-tech roster." *Business Week,* May 24, 1982, 68.

"The push for dominance in robotics gains momentum." *Business Week,* December 14, 1981, 108–109.

Reinhold, A. G., and Vanderburg, G. The autovision system. *Robotics Age,* Fall 1980, 22–28.

"Retraining displaced workers: Too little, too late?" *Business Week,* July 19, 1982, 178–185.

Robot Institute of America. *Worldwide robotics survey and directory.* Dearborn MI: Author, 1982.

"Robots make slow advance into electronic industry." *New Scientist,* January 22, 1981, 214.

"Robots to oust many workers by 1990 says labor secretary." *Christian Science Monitor,* February 2, 1982, 14.

Rosen, C. A., and Nitzan, D. Use of sensors in programmable automation. *Computer,* December 1977, 12–23.

"Russian robots run to catch up." *Business Week,* August 17, 1981, 120.

Saveriano, J. W. Industrial robots today and tomorrow. *Robotics Age,* Summer 1980, 4–17.

Shaiken, H. A robot is after your job. *New York Times,* September 3, 1980, A19.

"Smart, rugged robots are goal of project planned by MITI." *Electronics,* July 14, 1982, 79.

Smith, D. N., and Wilson, R. C. *Industrial robots: A Delphi forecast of markets and technology.* Dearborn MI: Society of Manufacturing Engineers, 1982.

Snyder, W. E., and Schott, J. Using optical shaft encoders. *Robotics Age,* Fall 1980, 2–11.

"The speedup in automation." *Business Week,* August 3, 1981, 58–67.

Spilhaus, A. Miniature mechanical marvels. *Technology Review,* January 1982, 51–57.

Sugarman, R. Blue collar robot. *IEEE Spectrum,* September 1980, 52–57.

Tanner, W. *Industrial robots.* Dearborn MI: Society of Manufacturing Engineers, 1978.

Tarvin, R. L. An off-line programming approach. *Robotics Today,* Summer 1981, 30–35.

Thompson, T. Robots for assembly. *Assembly Engineering,* July 1981, 32–36.

Tomizuka, M., Dornfeld, D., and Purcell, M. Applications of microcomputers to automatic weld quality control. *Journal of Dynamic Systems, Measurement and Control,* ASME, June 1980, 62–68.

U.S. Congress, Office of Technology Assessment. *U.S. industrial competitiveness.* Washington DC: Author, 1981.

Vasilash, G. S. The road to the automatic factory 1970–1981. *Manufacturing Engineer,* January 1982, 209–252.

"Video signal input." *Robotics Age,* March 1981, 2–11.

von Tiesenhausen, G., and Darboro, W. *Self-replicating systems—A systems engineering approach.* Marshall Space Flight Center Report 78304, 1981.

Vranish, J. The robotic deriveter. *Robotics Today,* Winter 1982, 24–28.

"Where the jobs are—and aren't. *Newsweek,* November 23, 1981, 88–90.

Wilson, K. R. Fiber optics: Practical vision for the robot. *Robotics Today,* Fall 1981, 31–32.

Winston, P. H. *Artificial intelligence.* Reading MA: Addison-Wesley, 1977.

ADDITIONAL READINGS

Aron, P. *Robots revisited: One year later.* Report 25, Daiwa Securities America, Inc., July 28, 1981.

Ayers, R., and Miller, S. Industrial robots on the line. *Technology Review,* May 1982, 35–46.

Cicela, J., et al. Future trends in manufacturing technology. *World-Wide Communication Journal,* July–August 1982, 138–154.

Dodd, G. C., and Rossol, L. *Computer vision and sensor-based robots.* New York: Plenum Press, 1979.

Friedrich, O. The robot revolution. *Time,* December 8, 1980, 72–83.

Kinnucan, P. How smart robots are becoming smarter. *High Technology,* October 1981, 32–40.

Kno, M. H. *Distributed computing on an experimental robot control system.* IEEE 1981 Proceedings of Applications of Minicomputers, 330–335.

Luria, D. UAW: We need technology. *Tooling and Production,* February 1982, 106–108.

Nagel, R. N., et al. Experiments in part acquisition using robot vision. *Robotics Today,* Winter 1981, 30–35.

"R & D spending surges, capital spending dies." *Business Week,* June 7, 1982, 16.

Robot Institute of America. *Worldwide robotics survey and directory.* Dearborn MI: Author, 1982.

"Robots join the labor force." *Business Week,* June 9, 1980, 97–103.

Romeo, G., and Camera, A. Robots for flexible assembly systems. *Robotics Today,* Fall 1980, 23–43.

Sanderson, R. J. *Industrial robots: A summary and forecast for manufacturing managers.* Naperville IL: Tech Tran Corp., 1982.

Schnapp, J. B. GM shakes up the auto industry. *Wall Street Journal,* May 26, 1982, 29.

Snyder, W. E. Micro-computer based path control. *Robotics Age,* Summer 1980, 6–15.

"Surge to robots is on." *Production,* December 1981, 104–106.

Takase, K., Paul, P. P., and Berg, E. J. A structured approach to robot programming and teaching. *IEEE Transactions on Systems, Man and Cybernetics,* April 1981, 274–289.

U.S. Congress, Office of Technology Assessment. *Exploratory workshop on the social impacts of robotics.* Washington: U.S. Government Printing Office, 1982.

Waters, C. R. There's a robot in your future. *INC,* June 1982, 64–74.

Index